Religion, Technology, and the Great and Little Divergences

History of Science and Medicine Library

VOLUME 32

Knowledge Infrastructure and Knowledge Economy

Edited by

Karel Davids, VU University, Amsterdam

Larry Stewart, University of Saskatchewan, Saskatoon

VOLUME 2

The titles published in this series are listed at brill.com/hsml

Religion, Technology, and the Great and Little Divergences

China and Europe Compared, c. 700–1800

By

Karel Davids

BRILL

LEIDEN • BOSTON
2013

Cover illustration: Frontispiece of Kaspar Schott, *Technica curiosa, sive mirabilia artis*, published in Nuremberg in 1664. The picture shows mathematical instruments and artefacts (including a copy of the Magdeburg hemispheres, top left) discussed in this book by Schott, a Jesuit scholar based in Würzburg, Germany. Courtesy Leiden University Library, The Netherlands, shelf mark 540 F 4.

Library of Congress Cataloging-in-Publication Data

Davids, C. A.
Religion, technology, and the great and little divergences : China and Europe compared, c. 700–1800 / by Karel Davids.
pages cm. — (History of science and medicine library, ISSN 1872-0684 volume 32. Knowledge infrastructure and knowledge economy ; volume 2)
Includes bibliographical references and index.
ISBN 978-90-04-23388-1 (hardback : alk. paper) — ISBN 978-90-04-23695-0 (e-book)
1. Technological innovations—Europe—History. 2. Technological innovations—China—History. 3. Technology—Religious aspects—Europe—History. 4. Technology—Religious aspects—China—History. 5. Technology—Social aspects—Europe—History. 6. Technology—Social aspects—China—History. 7. Cross-cultural studies—Europe. 8. Cross-cultural studies—China. I. Title.

T26.A1D38 2013
303.48'30951—dc23

2012032423

This publication has been typeset in the multilingual “Brill” typeface. With over 5,100 characters covering Latin, IPA, Greek, and Cyrillic, this typeface is especially suitable for use in the humanities. For more information, please see www.brill.com/brill-typeface.

ISSN 1872-0684
ISBN 978-90-04-23388-1 (hardback)
ISBN 978-90-04-23695-0 (e-book)

This book is printed on acid-free paper.

In memory of Suzanne Maria Davids-Hoefsloot

CONTENTS

LIST OF ILLUSTRATIONS

PREFACE

This book aims to connect four debates that are usually conducted separately: the debates on religion and technology, religion and economy, the Great Divergence between Asia and Europe and the Little Divergence within Europe. I believe each of these debates can hugely benefit from crossovers with the others.

I first became interested in exploring the links between these four discussions during the final phase of writing of my previous book, *The rise and decline of Dutch technological leadership. Technology, economy and culture in the Netherlands, 1350–1800,* which appeared in 2008. In the last chapter of that study, I argued that the particular way in which technology developed in the Dutch Republic in the late seventeenth and eighteenth centuries could be partly explained with reference to specific features of its religious-political context. The present book tackles a similar question on a much broader canvas – broader in a geographical and chronological, as well as in a thematic, sense. The purpose of this book is to throw light on the Great Divergence between China and Europe and the Little Divergence within Europe by analysing the relation between disparities in technological development and variations in religious contexts.

The emphasis in this book is on analysis and explanation. It does not pretend to give a full-blown overview of technological changes, material cultures or religious developments in China and Europe over 1,100 years. It attempts to answer the key question by means of a long-term, systematic comparison between China and Europe, based on a wide array of literature and a selected set of printed sources, which focuses on four important aspects: visions on the uses of nature, human capital formation, circulation of technical knowledge and technical innovation.

Parts of Chapters 3 and 4 have been presented as papers at conferences of the Global Economic History Network held in Wassenaar, Kaohsiung, Washington and London between 2004 and 2008, the Larry Epstein Memorial Conference in London in June 2008 and at sessions of the Second European Congress of World and Global History in Dresden in July 2008 and the Annual Meeting of the Society for the History of Technology in Lisbon in October 2008. I would like to thank all participants at these meetings for their inspiring and thought-provoking comments on earlier versions of my arguments.

Thanks to a grant from the Netherlands Organization for Scientific Research (NWO) and generous support from the History Department of Columbia University, New York, where my wife Marjolein 't Hart and I stayed for six months in the 2009–2010 academic year as Queen Wilhelmina Visiting Professors, I was able to complete a first version of the manuscript in the summer of 2010. The Columbia University Libraries provided invaluable resources for my research, and the pleasant surroundings of Maplewood, NJ, where we lived with our family, offered a relaxed atmosphere for writing. I am also grateful to Larry Stewart and two anonymous reviewers, whose careful and stimulating remarks helped to considerably improve my text. Finally, I would like to thank dr. Vivien Collingwood for English-language editing of the manuscript. This book is dedicated to my loving and wise mother Suzanne Davids-Hoefsloot, who passed away in August 2009.

Heemstede, March 2012

INTRODUCTION

The Great Divergence

One of the great debates in Global History concerns the origins of the 'Great Divergence'.[1] This term refers to the growing gap in economic performance between Europe and China (and by extension, Asia) since the late eighteenth century. A recurrent question in this ever-expanding debate concerns the extent to which the development of technology in Europe differed from that in China and how this difference, if it existed, could be explained.

The most extreme positions in this discussion have been taken by David Landes, on the one hand, and by André Gunder Frank, John Hobson and Jack Goody on the other. Landes claimed that from the Middle Ages onwards, Europe saw a continuous flow of inventions and accumulation of technical knowledge, which eventually (also aided by other factors) culminated in the Industrial Revolution in Britain.[2] Gunder Frank denied that there was such a thing as 'European' technology at all, let alone 'European technological superiority' before 1800. He argued that the 'very substantial' diffusion of technology and worldwide division of labour implied that technological development was a 'world economic process' rather than a regional or national one. There was, in other words, a set of shared human toolkits of technical knowledge, something that Edmund Burke III recently described as 'technological complexes'.[3] Gunder Frank insisted that even the technological advances of the Industrial Revolution, whose importance he admitted, should not be seen as purely European achievements.[4] John Hobson and Jack Goody likewise rejected the idea of a European lead in technological development before the Industrial Revolution.[5]

---

[1] The term came into vogue thanks to the book by Pomeranz, *Great Divergence*. For a review of the debate in the last ten years, see O'Brien, 'Ten years'.

[2] Landes, *Wealth and poverty*, 44–54, 187–193.

[3] Burke, 'Islam at the center', 167–168.

[4] Frank, *ReOrient*, 204, 285.

[5] Hobson, *Eastern origins*, especially 50–61 and 190–214, Goody, *Capitalism and modernity*, especially 19–20, 25–27.

Other scholars have taken a position somewhere between these extremes. Kenneth Pomeranz acknowledged that 'a surge in European technological inventiveness certainly [was] a necessary condition of the Industrial Revolution', but stressed the crucial role of 'coal and colonies' in lifting the constraints for sustained growth of per capita income.[6] Peer Vries, by contrast, argued that coal and colonies could only *become* relevant factors in the divergence between Europe and China because Western Europe, in contrast with China, had experienced 'a long process of continuing and self-sustaining invention and innovation' and by the eighteenth century had established a 'wide-ranging lead'.[7] Maxine Berg, Joel Mokyr and Prasannan Parthasarathi, too, have emphasized that technological development made the widespread adoption of coal possible. In their view, the divide between East and West essentially sprang from prior differences in technology. More specifically, Mokyr has related this development to the Scientific Revolution in the seventeenth century, which contributed to the rise of a new, unique environment for the creation of useful forms of knowledge, which he called the 'Industrial Enlightenment'. Parthasarathi has sought the origins of the exceptional technological change in England in a combination of external 'competitive pressures', 'ecological shortfalls' and a 'mercantile state'.[8]

Yet another perspective has been suggested by Jack Goldstone: on the one hand, he underscored the similarities between all societies in world history that experienced a period of economic, political and cultural 'efflorescence' (such as Song China, north-western Europe in the High Middle Ages, Golden Age Holland or High Qing China) and on the other hand, pointed to the unique emergence of 'engine science' – a reliance on engines for understanding the world – and its spread into popular culture in eighteenth-century England, which ushered in the Industrial Revolution.[9]

The differences of opinion on the development of technology thus largely revolve around three key issues: the question when and why Europe, rather than China, managed to develop a technological headstart on the rest of the world; the question whether Europe's technological lead resulted from a continued flow of inventions and accumulation of

[6] Pomeranz, *Great Divergence*, 68.
[7] Vries, 'Coal and colonies', 437.
[8] Berg, 'In pursuit of luxury', 90, Mokyr, *Gifts of Athena*, 34–42, *idem, Lever of riches*, 81, Parthasarathi, 'Great Divergence', 282, *idem, Why Europe grew rich*, 263.
[9] Goldstone, 'Efflorescences'.

technical knowledge or rather from a sudden, unique spurt in the process of technological change; and the question when and why the technological lead within Europe was focused within the Continent's north-western corner, and, more particularly, in England.

Until now, these issues have not been addressed in an interrelated, comparative way. While a fair amount of work has been done on aspects of technological development in China and Europe before 1800, systematic comparisons of variations in technological change between China and Europe and within Europe itself have not been made.[10] Although the relevance of technology for the explanation of differences in economic performance is clearly acknowledged, as the discussion above demonstrates, the bulk of recent research on the origins of the Great Divergence concentrates on measuring and analysing data on variables like wages, prices, production, consumption, exports, imports, taxation and government expenditure, rather than on studying contexts and mechanisms of technological development. Technology is being regarded as given, instead of being treated as an explanandum. Factor endowments and factor prices figure much more prominently than the supply of 'useful' knowledge.[11] Moreover, although there have been a few state-of-the-art surveys of technological development in world history that cover both China and Europe,[12] rather than making direct comparisons between these two regions, their trajectories are presented as entirely separate stories. A recent survey on the Industrial Revolution in England in global perspective, for instance, contained an analysis of technological change in England (and to some extent in France), but not of technology in China.[13]

Variations in technological development in China and Europe are the subject matter of this book. The key problem is: how can variations in the development of technology between Europe and China and within Europe itself before c.1800 be explained? This calls for a systematic, comparative approach. The angle from which these comparisons will be made is that of the relationship between technology and religion. I suggest that research into variations in technological change can be fruitfully

---

[10] O'Brien, 'Needham question updated', 10.

[11] Pomeranz, *Great Divergence*, Vries, *Via Peking back to Manchester*, Parthasarathi, 'Rethinking wages', Broadberry and Gupta, 'The early modern Great Divergence', Van Zanden, *Long road to the Industrial Revolution*, Lal, *Unintended consequences*, Rosenthal and Wong, *Before and beyond divergence*.

[12] Mokyr, *Lever of riches*, McClellan and Dorn, *Science and technology*; Friedel, *Culture of improvement*, deals extensively with Europe but barely with China.

[13] Allen, *British Industrial Revolution*.

combined with ideas, conjectures and insights produced in debates on the impact of religious factors on technical and economic change, which started long before the issue of the 'Great Divergence' moved to the forefront of historical discussion. David Landes noted that various scholars have sought the explanation for the European 'invention of invention' in a particular set of religious values, such as the 'Judeo-Christian' 'respect for manual labor', 'the subordination of nature to man' and 'sense of linear time', although he himself preferred to seek the ultimate cause in market forces.[14] Rolf Sieferle likewise included religious factors among the relevant circumstances that made the eventual industrial transformation in Europe possible.[15] And Joel Mokyr has observed that 'religion, technology and society were often intertwined in ways that defy easy generalization but that had an evident effect on technological creativity'.[16]

Technological Change in China and Europe between c. 700 and 1800

What do we actually know about technological change in China and Europe from a comparative perspective? What are the answers to the 'when' and 'whether' questions identified above? Whenever somebody tries to make a comparative statement about technological development in China and Europe before 1800, three emblematic examples invariably crop up: the outburst of ocean voyaging under Zheng He, the failure of the Macartney Embassy and the unfulfilled promise of Su Song's water clock.

Ever since his dazzling appearance in volume IV of Joseph Needham's *Science and civilisation in China*,[17] Admiral Zheng He has been routinely wheeled out as the prime exhibit of Chinese technological superiority at the beginning of the fifteenth century. The seven voyages of 'exploration' and 'friendship' conducted by huge fleets of 'treasure ships' and smaller vessels under the command of Zheng He and other eunuchs between 1405 and 1433, which sailed to Southeast Asia and into the Indian Ocean as far as East Africa and the Arabian Peninsula, are often regarded as proof that the Chinese were then already capable of performing technical feats

---

[14] Landes, *Wealth and poverty*, 58–59; O'Brien, 'Needham question updated' likewise draws attention to the differences in 'cultures and cosmologies for innovation', including religious traditions, 10, 13–22.

[15] Sieferle, *Sonderweg*, 50.

[16] Mokyr, *Lever of riches*, 172.

[17] Needham, *Science and civilisation*, vol. IV part 3, 487–535, Finlay, 'China, the West', 268–269.

that Europeans would be unable to match for a long time to come. Zheng He's exploits have even been equated with the American manned voyages to the moon in the 1970s. Compared to the mighty Chinese junks that crossed the Western Ocean, the puny ships captained by Columbus and Vasco da Gama paled into insignificance.[18] Sceptics, by contrast, have stressed another aspect of these expeditions: the complete withdrawal of imperial support for such undertakings after the 1430s and their lack of follow-up. This reversal is taken is a sign that the Chinese economy turned in upon itself.[19]

The British embassy to China led by Viscount George Macartney in 1792–1794 to facilitate access to the Chinese markets for British traders has become a symbol both of 'a failure of understanding by the west of China's achievements' and of China's failure 'to recognize the recent technological progress of the west'.[20] The goods that Macartney brought as gifts to Beijing on behalf of King George did not impress the Imperial Court: 'We have never set much store on strange or ingenious objects, nor do we need any more of your country's manufactures', the Emperor Qianglong plainly stated in his reply to the British King. In the end, the British did not receive any commercial concessions at all.[21]

The huge, water-driven astronomical clock designed by the polymath Su Song on the order of Emperor Zhezong, which was installed in the imperial capital Kaifeng in 1094 and which is said to have kept time more accurately than any other device available at that time, has been hailed by some sinologists and historians (Needham foremost) as a dramatic piece of evidence that China was far ahead of the West in this critical area of technological achievement.[22] Others, however, have preferred to view Su Song's clock merely as 'a magnificent dead end'. Not only was the intricate device dismantled after the conquest of North China by the Jürchen in 1127, but it also failed to spawn further advances in mechanical clock making, and faded almost completely from memory under the Ming and Qing. Su Song's clock may have been a fantastic feat of engineering, but it remained, at the end of the day, a white elephant.[23]

---

18 Goldstone, 'The rise of the West', 176–177.

19 See e.g. Landes, *Wealth and poverty*, 93–98.

20 Berg, 'Macartney embassy', 270–275.

21 Berg, 'Macartney embassy', 278, 285–286, Landes, *Revolution in time*, 49.

22 Needham et al., *Heavenly clockwork, passim*, Needham, *Science and civilisation*, vol. IV part 2, 446–465.

23 Pagani, *"Eastern magnificence"*, 5–13, Landes, *Revolution in time*, 17–36, Cipolla, *Clocks and culture*, 40.

None of these readings can entirely withstand closer scrutiny. Zheng He's expeditions were in reality more a show of force than a series of voyages of exploration. The fleets of 'treasure ships', which were incidentally much smaller than legend has made them out to be, did not discover any new sea routes or any unknown worlds overseas. The Chinese obtained their knowledge on Africa from Arab and Persian sources. Thanks to these sources, China under the Yuan dynasty already had 'a surprisingly detailed body of knowledge on the geography of Africa and Europe and the lands and seas in between'.[24] The voyages led by Zheng He 'were essentially an urbane but systematic tour of inspection of the known world', Needham acknowledges.[25] In fact, Zheng He's fleets look much more like Teddy Roosevelt's Great White Fleet, which circumnavigated the earth in 1907–1909, than the Apollo missions to the moon six decades later. On the other hand, the end of the great expeditions did not spell the end of China's feats as a seafaring nation. Chinese ships continued to sail far and wide across the Asian seas throughout the early modern period.[26] The fate of Macartney's embassy, in turn, does not tell us anything about Chinese or British superiority in technology at the end of the eighteenth century. One of the reasons why Chinese court officials showed a lack of interest in the ambassador's gifts was that most of the goods offered, such as clocks and mathematical and scientific instruments, had already been brought in large numbers from Europe to China, or had been constructed by Europeans in China, in the preceding 150 years. Consumer goods or industrial products that *did* represent recent British advances in technology were, despite the pleas of leading manufacturers such as Matthew Boulton, hardly included in the package of gifts that eventually arrived in Beijing.[27] And as far as clocks are concerned: although Su Song's device remained unique in its kind, clock-making in China evolved further during the Yuan and early Ming periods.[28]

The different interpretations of such events as the voyages of Zheng He, the Macartney Embassy and the construction (and deconstruction) of Su Song's clock reflect deep-seated assumptions about the relative levels or rates of technological development in China and Europe. On the one

[24] Allsen, *Culture and conquest*, 114.

[25] Needham, *Science and civilisation*, vol. IV part 3, 499, 529, Wake, 'Myth', 74–75, Wade, 'Zheng He's voyages', Finlay, 'China, the West', 295–299.

[26] Waley-Cohen, *Sextants of Beijing*, 105–121, Pagani, *"Eastern magnificence"*, 115–170, Curtis, *Glass exchange*, 36–37, 41–53, 60–61, 113–117.

[27] Berg, 'Macartney embassy', 279–285, 287.

[28] Pagani, *"Eastern magnificence"*, 13–14, 58–98.

hand, we encounter confident statements such as that made by Francesca Bray, that 'historians who have studied Chinese technology agree that after an initial flowering up to about 1400, during which China surpassed Europe in productive capacity and inventiveness, China fell into a period of stagnation and decline';[29] and on the other hand, equally bold claims such as that advanced by Michael Adas, building on work by historians of the *Annales* School, to the effect that already 'by the fifteenth century the peoples of western Europe possessed an advantage of three or four to one over the Chinese in per capita capacity to tap animate and inanimate sources of power'.[30] Such contentions are no more than guesses, however. What 'historians who have studied Chinese technology' (especially Joseph Needham and his companions) are claiming in comparisons with Europe is often based more on an intimate acquaintance with China than on knowledge of European history.[31] Comparative data on 'productive capacity' in the fifteenth century are lacking. The extant data on sources of power are scarce and fraught with difficulties of interpretation.[32] While Paolo Malanima is probably right in stating that by the mid-eighteenth century, the economy of pre-industrial Europe was more energy-intensive than the Chinese economy and, in comparison to China, relied more on energy from wood and coal and less on energy provided by human muscles, it is not clear exactly *when* this difference emerged and at what rate it increased. Moreover, even in the middle of the eighteenth century, shares of other energy sources, such as wind and waterpower, in total energy production in Europe and China hardly differed.[33]

In fact, in terms of the sheer variety of machines used, China and Europe before 1800 look remarkably similar – up to a point. For China, Needham describes a wide range of what he – following Lewis Mumford – calls 'eotechnic machinery': wooden or bamboo devices with a number of (often revolving or spinning) components to carry out certain productive functions, powered by animals, humans or water. Among these machines were querns, animal-powered grinding mills, hand-driven chain pumps,

---

29 Bray, *Technology and gender*, 7.

30 Adas, *Machines*, 21, referring to guesstimates by e.g. Pierre Chaunu and Fernand Braudel in *Civilisation matérielle et capitalisme*; although Adas acknowledges the 'poor quality of the data', he nevertheless seems to accept their validity.

31 A glaring example of Needham's lack of familiarity with European history is given in Davids, 'River control'.

32 See e.g. Malanima, 'Energy crisis', 105, 107 note 24, Malanima, *Pre-modern European economy*, 88–93, Unger, 'Energy sources', Davids, 'Innovations', esp. 273–275, 288–289.

33 Malanima, *Energia*, 119–125, Malanima, 'Energy crisis'.

scoop-wheels moved by humans or animals for lifting water into fields, winches for mine-shafts, human-driven winnowing-fans, water-powered bellows, water-driven multi-spindle spinning frames, man-powered looms and vertical two-roller sugar cane crushers powered by animals.[34] The variety of machines used in Europe between the late fourteenth and early nineteenth centuries was at least as large as in China. All the different types of machines and mechanical devices described in volume IV of *Science and civilisation in China* can also be found in late medieval and early modern Europe, although they are not always made of the same sort of materials.

The mundane reality is that China, *pace* Needham, was not the only cradle of invention in the world before 1500 and neither was Europe, *pace* Landes, Adas and the *Annales* School, uniformly more technologically advanced than China in the early modern period. Between c.700 and 1800, both China and Europe went through long periods of incremental technological change,[35] interspersed with outbursts of accelerated creativity. The fact that a revolutionary technological breakthrough was eventually achieved in Europe and not in China should not lead us to exaggerate the differences in the general *level* or *rate* of technological change in Europe and China before the end of the eighteenth century. Yet, even if the general level and rate of technological change in China and Europe before the end of the eighteenth century did not vary largely, there were differences in the *nature* of technological change.

Benjamin Elman's evaluation gives an idea of these differences in nature:

> arguably, by 1600 Europe was ahead of Asia in producing basic machines such as clocks, screws, levers and pulleys that would be applied increasingly to the mechanization of agricultural and industrial production. In the seventeenth and eighteenth centuries, however, Europeans still sought the technological secrets for silk production, textile weaving, porcelain making, and large-scale tea production from the Chinese.[36]

Even this more carefully worded evaluation is still somewhat unbalanced, however, because the second part of the statement leaves out the reverse side of the story. What about the technological secrets that the Chinese still sought from the Europeans? The demand for the skills of

---

[34] Needham, *Science and civilisation*, vol. IV part 2, *passim*, esp. table 56.

[35] Mokyr, *Lever of riches*, 7, 57–58, 209–219.

[36] Elman, *On their own terms*, xxxi.

Jesuit missionaries from Europe in the eighteenth century is one clue to this. At the court of the Qing, Jesuits were not only highly valued for their superior knowledge of astronomy and mathematics and their expertise in building machines and in making clocks, automata and scientific instruments, but also for their abilities in a wide range of other fields. Notably, Jesuit missionaries supervised the founding of cannon, conducted surveys of the empire, produced maps, made oil paintings, led operations in glass-making, designed fountains, gardens and palaces and introduced new plants and medicines.[37] And this yardstick does not even include those 'secrets' that the Chinese did not know or care about, such as the secret of making cheese from the milk of dairy animals.[38]

Variations in technological evolution between China and Europe concerned the degree of development, or reversals in development, of particular techniques or artefacts, as well as a particular sector's technological advance. Technological evolution in China primarily differed from developments in Europe in the sense that particular techniques or artefacts that were common to both regions of Eurasia ultimately evolved to a lesser extent in China than in Europe. The potential of specific innovations was not always realized as fully in China as in Europe, even if they *did* come into widespread use.[39] Clocks, cannon, the magnetic compass and printing by movable type are perhaps the best-known examples of this phenomenon, but they are by no means the only ones. Other examples can be found in sugar production, metalworking and mining. The vertical two-roller sugar cane crusher, powered by animals, entered China in the seventeenth century at about the same time as European plantation colonies in the Americas, but did not evolve into the even more productive animal-driven or water-driven three-roller type.[40] Hans Ulrich Vogel has drawn attention to the fact that from the sixteenth and seventeenth centuries onwards, copper smelting in Europe became much more mechanized than in China, even though the necessary knowledge was to some extent available in China as well.[41] Peter Golas pointed out that until the twentieth century, Chinese mining largely remained a 'small-scale, primitive'

---

37 Waley-Cohen, *Sextants of Beijing*, 105–121, Curtis, *Glass exchange*, 36–37, 41–53, 60–61, 113–117, Musillo, 'Reconciling two careers'.

38 Needham, *Science and civilisation*, vol. VI part 5, 328–329: 'there is no evidence that cheese [from milk from dairy animals] was ever made in China'.

39 Mokyr, *Lever of riches*, 218.

40 Mazumbar, *Sugar and society*, 138–159, 178–180, Sabban, 'Industrie sucrière', 824–831, Needham, *Science and civilisation*, vol. VI part 3, 335.

41 Vogel, 'Kupfererzeugung', 45, *idem*, 'Mining industry', 172.

affair. Due to the abundance of cheap labour, labour-saving techniques were hardly employed and mechanization was almost absent. Mining in China was 'crude in the sense that it was carried on largely by human labour using simple tools'.[42]

Yet another illustration is the case of windmills. Historians of technology in China have pointed out that wind-powered water-lifting devices were described in a book dating from the twelfth century and that windmills became quite widespread in coastal areas in Southeast China under the Ming. At the time, wind power was used for grinding corn as well as for raising water.[43] In Europe, however, wind power came to be employed for a much greater variety of purposes than in China. In Europe, the power of wind was not only harnessed for grinding corn or lifting water, but from the sixteenth century onwards, it was also used for all sorts of industrial processes, such as fulling, oil pressing, hemp crushing, timber sawing, hulling, cannon-boring, grinding mortar or cutting rags as raw material for paper. Northwest France, Flanders and Holland in particular were centres of innovation for these new applications of wind power. By the early nineteenth century, thousands of windmills for all sorts of functions were in operation across Europe.[44] Technology *in use* in China and Europe might thus have differed to a greater extent than a survey of mere technological *potential* suggests.

Moreover, after c.1000, regression in technological development seems to have occurred more frequently in China than in Europe. Here I am using the term 'regression' not in the 'hard', strict sense in which it has appeared in the economic historical literature of the past few years – as 'lengthy periods during which important knowledge was lost'[45] – but in the 'softer', more flexible sense of a prolonged, large-scale shift from a technology of some complexity to a more simple technique. Aside from the special cases of Tasmania or Easter Island,[46] actual, complete 'loss' of existing technology in pre-industrial economies in entire territories or continents has rarely been observed. All known cases of technological regression in Europe in the 'soft' sense occurred in the Early Middle Ages,

---

[42] Needham, *Science and civilisation*, vol. V part 13, 16–17, 429–434.

[43] Needham, *Science and civilisation*, vol. IV part 2, 558, Zhang, 'Ancient Chinese windmills'.

[44] Davids, 'Innovations in windmill technology', 286–287, 290–291.

[45] Aiyar, Dalgaard and Moav, 'Technological progress and regress', 126, 136–137, Persson, *Economic history*, 26–28.

[46] Used as prime exhibits of technological regress in 'pre-industrial economies' in Aiyar, Dalgaard and Moav, 'Technological progress and regress', 129, 140–141.

namely during or shortly after the disintegration of the Roman Empire in the West. Hydraulic mining techniques in northwest Spain were abandoned after the third century AD. The use of the potter's wheel disappeared in Britain after the Roman armies had left the island. From the fifth century onwards, fired brick (and concrete) was not used in building north of the Alps for a long time.[47] In hydraulic technology, sophisticated water systems from Roman times were neglected or abandoned in many areas in Europe during the Early Middle Ages. In many places, aqueducts fell into disrepair. The technique of making 'pozzolana cement', which had the advantage of hardening under water, passed into disuse.[48] Once such technologies had been recovered, however, there was never a turning back to older, simpler modes of water provision. Regarding inanimate sources of energy, recurrent changes in the relative frequency of water, wind and horses as sources of power did occur, but after the High Middle Ages, there was never a large-scale shift back from water or animal power to human power in Europe.[49]

But these things *did* happen more than once in China after the Song. In farming in Jiangnan, ox tilling replaced human-powered tilling during the Tang, but this substitution was reversed during the Ming. In the late Ming, 'ox tilling was rather rare'. During the early Qing, however, the substitution occurred yet again, but the other way around: tilling by oxen replaced ploughing by men.[50] Water-powered multiple spinning frames seem to have gone out of use in the fourteenth century.[51] Under the Qing, waterpower for moving bellows in blast furnaces in the iron industry was frequently replaced by human labour.[52] Peter Golas has stated that mining technology showed 'sluggish development and even occasional retrogression'. The process of producing copper from vitriol waters, for instance, which reached a high degree of sophistication under the Song, passed largely into disuse thereafter.[53]

These variations in the degree of development of techniques and artefacts and in the relative frequency of reversals seem to support a claim advanced by Bozhong Li, following a distinction made by Francesca Bray,

---

47 Wilson, 'Machines', 32.

48 Magnusson, *Water technology*, 4, Aiyar, Dalgaard and Moav, 'Technological progress and regress', 127–128.

49 See for example Langdon, *Mills in the medieval economy*, 34–40.

50 Li, *Agricultural development*, 43–44.

51 Needham, *Science and civilisation*, vol. V part 9, 236.

52 Wagner, *Iron and steel, idem, Traditional Chinese iron industry*, 31, 45.

53 Needham, *Science and civilisation*, vol. V part 13, 370–386, 428.

that technologies in Europe were 'mechanical' and those in China were 'skill-oriented'. 'Mechanical' technologies 'favour the development of equipment and machinery as a substitute for human labour', while 'skill-oriented technologies' 'tend towards the development and intensive use of human skills, both practical and managerial'.[54] Yet, this distinction only holds up to a point. Although Europe indeed followed a more *consistent* trajectory than China in 'mechanical' technologies, the economically most advanced parts of Europe, such as the Netherlands, also continued to make great strides in 'skill-oriented' technologies throughout the early modern period. The technological leadership of the Dutch Republic in Europe in the seventeenth and early eighteenth centuries was as much based on supremacy in 'skill-oriented' technologies as mastery of 'mechanical' technologies.[55] On the other hand, historians of technology in China, including Bozhong Li himself, have stressed in recent years that, contrary to influential views, such as that of Mark Elvin, Qing China *did* see the adoption of labour-saving improvements.[56] Bozhong Li has argued, for example, that multi-spindle spinning machines invented in the fourteenth century for spinning hemp were adapted for spinning cotton in the eighteenth century and remained in use in Songjiang (Jiangnan) until the mid-twentieth century.[57] The distinction between 'mechanical' and 'skill-oriented' thus does not coincide with the distinction between 'the West' and 'East Asia' as neatly as Li claims – and for this reason, neither can attention to advances in 'mechanical' technologies easily be labelled 'Eurocentric'.[58]

Equally noteworthy is the difference in emphasis by sector of technological advance that emerged between China and Europe after about 1500. In contrast with Elvin's thesis that late Ming and Qing China suffered from 'a relative technological standstill' due to a combination of high agricultural productivity, nearly static markets and cheapening labour ('the high-level equilibrium trap'),[59] in recent years, historians of China have argued that Qing China witnessed significant technological advances, albeit mainly of an incremental nature and with an emphasis on the diffusion of knowledge rather than innovation. What is striking, though, is that these historians nearly always lay stress on changes in the *agricultural* sector, not on

[54] Li, *Agricultural development*, 167–168, referring to Bray, *Rice economies*, 115, 156.
[55] Davids, *Rise and decline*, esp. chapter 3.
[56] Li, *Agricultural development*, 54, Mazumbar, *Sugar and society*, 188, 190, Sabban, 'Industrie sucrière', 857, Elvin, *Pattern*, 215, 314.
[57] Li, 'Involution', 389, cf. Chao, *Development*, 59, 63, Zurndorfer, 'Cotton textiles', 51.
[58] Li, *Agricultural development*, 168.
[59] Elvin, *Pattern*, 301, 312–315.

innovations in urban industries, maritime transport or extraction of raw materials. When they want to make the point that Qing China *did* see some technological change, they describe the diffusion of new crops (such as sweet potatoes or maize) and of new varieties of rice, the dissemination of new ways of tilling the soil and the spread of sundry improvements in cropping systems, fertilizer use or sugar processing techniques.[60] Cotton cloth production also underwent technical changes in Ming times, but, apart from some specialist, city-based workshops, it remained almost entirely an activity pursued by rural households.[61] Porcelain manufacture was an urban affair, concentrated in the town of Jingdezhen in Jiangnan. Although its products reached wider markets and enjoyed a better reputation in Ming and Qing times than ever before, the rate of innovation in this urban industry in this period was actually lower than under the Song and Yuan. The most important improvements in the mixing of materials and in kiln technology were made before the sixteenth century. The main innovation in Ming times was the use of new colours in enamelling, which may have been inspired by contact with Europe.[62] A historian of the porcelain industry in Jingdezhen emphasizes that 'technical excellence had reached a point long before the Ming at which it was to remain until the twentieth century...Jingdezhen relied on the *tradition* of manufacturing rather than on *innovation* to supply its high quality wares'.[63] In raw material extraction, the only sector to see continued technological change until the nineteenth century was salt production in Sichuan. Since the eleventh century, salt in Sichuan had been extracted by a technique of deep drilling with heavy iron percussion drills. In the course of time, and especially in the eighteenth and early nineteenth centuries, these mechanical, man-powered drills were improved in such a way that they could reach ever greater depths. The deepest boreholes in the late sixteenth century were 300 metres deep, and in the 1830s they were more than three times as deep.[64]

The scope of technological change in early modern Europe was broader than in Late Imperial China. It took place both in the agricultural sector and in urban industries, as well as in other economic sectors such as

---

[60] Li, *Agricultural development*, 39–54, Mazumbar, *Sugar and society*, 138–191, Deng, 'Critical survey', 15.

[61] Zurndorfer, 'Cotton textiles', 53–56, 62, Li, 'Involution', 389–390.

[62] Needham, *Science and civilisation*, vol. V part 12, 229–233, 314–359, 366–378, 443, 552–560, 619, 638–643, Gerritsen, 'Ceramics', 165–171, 173, 175–177.

[63] Dillon, 'Jingdezhen', 38–39.

[64] Vogel, 'Salzgewerbe', 92–98.

mining and maritime transport. In all of these sectors, innovation continued throughout the sixteenth, seventeenth and eighteenth centuries, as has amply been shown for the Netherlands, England and Central Europe.[65] Eventually, it was Europe, not China, where the transition to a modern industrial economy began and sustained economic growth took off. The beginning of modern economic growth is nowadays widely seen as a result of a dramatic shift in the energy basis, which allowed an unprecedented rise in energy consumption and a huge increase in productivity. This change in the energy basis, which essentially revolved around a large-scale transition to the use of energy sources stocked *below* the surface of the earth, namely fossil fuels, implied a fundamental transformation in the relation between human societies and the natural environment. It has variously been described as the rise of a new 'social-metabolical regime' (Sieferle), as a shift from an 'organic economy' (Wrigley) to a 'mineral-based energy economy' or as a transition to a different energy system based on different carriers of energy (Malanima).[66] By shifting to this new energy regime, from the late eighteenth century onwards, Europe truly set out on an entirely different path of economic and technological development than China and the rest of the world.

Where did these variations in technological change between China and Europe come from? This brings us to the 'why' questions that form the heart of this book. This study does not focus on proximate causes, such as relative factor prices, but underlying factors of technological development. The book aims to contribute to this discussion by approaching the issue in a more systematic, comparative way than has hitherto been attempted. The angle from which these comparisons will be made is the relationship between technology and religion; and this requires, first of all, a survey of the debate on religion, the economy and technology.

---

[65] See e.g. Davids, *Rise and decline*, esp. chapter 3, Vogel, 'Mining industry', 170–172, Vogel and Theisen-Vogel, 'Kupfererzeugung', 41–45, Allen, *British Industrial Revolution*, 156–163.

[66] Sieferle, *Unterirdische Walt*, esp. 144–171, Sieferle, *Sonderweg*, 30, 33, 38, Wrigley, *Continuity*, 17–33, Malanima, 'Energy crisis'.

Religion, the Economy and Technology Before 1800: Different Views

Weberian Themes and the Little Divergence

More than a century ago, Max Weber famously advanced his thesis on the relationship between the growth of 'the spirit' of capitalism in the Late Middle Ages and early modern period and the 'Protestant ethic'. He suggested that distinctive traits of what he called 'ascetic Protestantism' (notably Calvinism), such as diligence in one's spiritual and vocational calling, inner-worldly asceticism and the urge to make the utmost use of one's time, were eminently favourable for the emergence of a modern capitalist mentality. Once this 'spirit' was formed, Weber assumed, it could also spread to members of other denominations, including Catholics or Buddhists. However, he did not elaborate on the conditions under which a diffusion of the ascetic Protestant model could occur.[67]

Although Max Weber's basic contention has been characterized as 'arguably obtuse, practically impossible to confirm or demonstrate, and remote from twentieth-century concerns',[68] more than a century after it was first formulated, his thesis is still a frequent subject of debate among social scientists. Unlike most other time-honoured theories in sociology, Weber's thesis has to date survived all attempts at elimination by refutation, trivialization or outright neglect. In 2005, the editors of a collective volume dedicated to the first centennial of his thesis triumphantly concluded from a survey of recent literature that the subject was still very much alive, even though critics of various provenance had more than once declared it dead and buried.[69] One of the reasons for its endurance is, no doubt, as Jack Barbalet has noted, that 'every undergraduate student enrolled on a sociology course [is still] expected to know the *Protestant Ethic*'.[70] Historians, too, often give a dutiful nod to Weber (cloaked in some critical remarks, of course) before proceeding to the matter at hand.[71]

While Weber himself never explicitly discussed the possibility of a relationship between the 'Protestant ethic' and technological change (probably because his main field of interest was the development of values,

---

67 Weber, *Religion und Gesellschaft*, 26–28, 179, 281.

68 Barbalet, *Weber, passion, and profits*, 2.

69 Swatos and Kaelber (eds.), *The Protestant Ethic turns 100*.

70 Barbalet, *Weber, passion, and profits*, 14.

71 See e.g. Pomeranz, *Great Divergence*, 77, 111, 167, 177, Mokyr, *Gifts of Athena*, 42, Parthasarathi, *Why the West grew rich*, 80–82.

institutions and organizations),[72] in *Das technische Zeitalter* in 1928, theologian and art historian Hanns Lilje dared to make a more explicit claim. Technology arose on the fertile soil of the Renaissance, he argued. It was the spirit of the Renaissance which released humans from their close, medieval bonds to the natural world and enabled them, 'with the full freedom of creative technical thinking', to undertake the attempt to subjugate nature. Yet, Lilje contended, while the Renaissance created this freedom, the Reformation directed its use. 'Wie das moderne Wirtschaftsleben vom Kalvinismus starke Antriebe empfangen had' (Lilje evidently knew his Weber), 'so ist auch in der Technik protestantischer Geist verborgen wirksam gebleben'. The newly acquired freedom was used to oblige God and humanity. Technicians who understood their job properly strove to employ their powers to serve their community, their fellow men and their Creator.[73]

An American sociologist, Isidor Thorner, stated in 1952 that Weber's thesis could actually explain differences in inventiveness between countries. Assuming that 'affective neutrality' (the ability 'to suspend judgment in emotionally important situations') was relatively more prominent among 'ascetic Protestants' than among Catholics, Thorner reasoned that ascetic Protestantism was therefore apparently more congenial to technological advance than Catholicism. Taking issue with Pitirim Sorokin's confident assertions to the contrary, Thorner insisted that ascetic Protestantism had a more 'stimulating influence... on science and technology' than Catholicism, as the 'productivity of discoveries and inventions' of countries influenced by ascetic Protestantism, such as in England, Holland or Pietist Germany – according to his statistics – was higher than that of Catholic 'strongholds' such as Italy, Portugal or Spain. Moreover, a substantial proportion of 'inventiveness' in Catholic countries could in fact be ascribed to people with an ascetic Protestant background. Max Weber had been vindicated again, Thorner triumphantly declared.[74]

For Charles Webster (followed by David Noble), it was not the supposed 'ascetic' quality of Protestantism per se, but rather the zealous, unremitting drive of the Puritans to bring about the restoration of Paradise by means of a growing knowledge and manipulation of nature that gave a powerful impetus to technological advance in England. The

[72] Cf. Elvin, 'Why China failed', 383, and Weber, *Wirtschaft und Gesellschaft*, 32–33.
[73] Lilje, *Das technische Zeitalter*, 86–87.
[74] Thorner, 'Ascetic Protestantism', 25–33.

'unprecedented millenarian milieu' that came into being in the late sixteenth and seventeenth centuries, Noble argued, 'decisively and indelibly shaped the dynamic Western conception of technology'.[75] Henry Heller stated that 'the drive towards technological improvement', which was evident in France in the second half of the sixteenth century, 'was strongly informed by Calvinist religious assumptions'; without, incidentally, specifying how this influence took shape. According to Heller, nearly all of the leading writers on technology (and natural philosophy, for that matter) were 'inspired by evangelical ideas'. 'Sympathy for the teachings of Calvin' and 'a desire to reform or create knowledge and improve society' were thus, in his view, closely linked.[76]

Recently, Philip Gorski has suggested a revised version of Weber's thesis, which comes down to the claim that the 'Protestant Reformation' was one of the underlying causes of 'the Little Divergence' in early modern Europe. Gorski conceives the 'Protestant Reformation' not in the specific Weberian sense of 'ascetic Protestantism', nor as 'a discrete event spanning a few years', but as 'a series of religious reform movements that lasted a full century and led not only to the break-up of the Catholic Church but to schisms within the Protestant camp'. This concept thus includes the rise of Lutheranism as well as the spread of Calvinism and various other Protestant reform movements. By 'Little Divergence' Gorski means the growing economic divergence in the seventeenth and eighteenth centuries between ascendant areas in north-western Europe, notably Holland and England, and less dynamic parts of the Continent, such as Spain, France and the Italian peninsula.[77]

While Gorski's thesis, like Weber's, assumes a connection between economic change and particular religious variables, the causal mechanisms involved are different. The connection does not reside in religious ideas or institutions, but in the impact of the Protestant Reformation on economic resources and incentives and political and social institutions that are favourable to economic growth, in particular the accumulation of human capital, the supply of labour, the structure of landholding and the rise of more responsible and more predictable forms of government.[78] And because some of these factors also directly or indirectly influence

---

[75] Webster, *The Great Instauration*, 325–327, 369–384, 506–507, Noble, *The religion of technology*, 39–40.

[76] Heller, *Labour*, 118.

[77] Gorski, 'The Little Divergence', 168, 187–188.

[78] Gorski, 'The Little Divergence', 167–169.

the context for technological innovation, Gorski's claim is relevant for the question of the impact of religious factors on technological innovation.

Economic historians have also adopted the term 'Little Divergence' to denote the increasing gap in per capita income between north-western Europe and the Mediterranean and East-Central Europe in the early modern period.[79] In contrast with the 'Great Divergence', the notion of the 'Little Divergence' has up to now only been used in an intra-European context. Although China also experienced regional disparities in economic development and religious practices, and the Yangzi is sometimes taken as a demarcation line between religious traditions,[80] there are as yet no signs of a debate about possible *relations* between economic and religious differences. A counterpart to the argument on the 'Little Divergence' in Europe does not yet exist. Relations between economic and religious differences between, say, North China and Jiangnan are not discussed in the same way as those between Castile and the Dutch Republic.

Latin Christendom and China

Next to the ever-swelling literature on the Weber thesis, which spawned an entire offshoot in the study of technological change, the role of religion in technological advance as such has also become a subject of a wide-ranging debate. Initially at least, this debate largely focused on the possible influence of Christian ideas and values, and more specifically the influence of the ideas and values of Latin Christendom, on the development of technology.

In its most general form, the argument runs that Christianity and/or the organized Church paved the way for technological advance by creating a sharp distinction between man and nature. Because Christianity is said to place man above the rest of physical creation, while denying rivers, trees or hills their own 'spirits' and refusing to accept the view of the cosmos as a living organism, it is supposed to have laid the basis for a more domineering and conquering attitude on the part of humans towards the natural world. It was easier to manipulate nature once nature had lost its soul.

---

[79] Broadberry and O'Rourke, 'Introduction', 3, Van Bavel, De Moor and Van Zanden, 'Institutional organization', 9.

[80] Skinner (ed.), *City in Late Imperial China*, Skinner, 'Presidential address', Pomeranz, *Great Divergence*, 226, 234, Ter Haar, 'Buddhist-inspired options', 95, Naquin and Yü, 'Introduction'.

This kind of argument first appeared in studies by R.J. Forbes, S. Sambursky and others published in the 1950s.[81] Against the background of rising ecological concerns in the late sixties and early seventies, it received a more radical twist in publications by Lynn White jr. and Carl Amery. In an original and highly provocative article published in *Science* in 1967, Lynn White asserted that the historical roots of the imminent 'ecological crisis' could be located in the influence of Christian (and especially: *Latin* Christian) attitudes towards the natural environment, which had led to a ruthless exploitation of the physical environment by technological means. In his view, the 'increasing disruption of the global environment' was the 'product of a dynamic of technology and science' resulting from 'Christian attitudes towards man's relation to nature which were almost universally held not only by Christians and neo-Christians but also by those who fondly regard themselves as post-Christians'. After all, 'modern' technology and science was, according to White, 'distinctively Occidental', and the 'superiority' of the West in technology was already achieved before the end of the Middle Ages, when the Christian 'dogma' of 'rightful mastery over nature' had been firmly established. Christianity therefore bore a 'huge burden of guilt' for the ecological predicament in which humankind found itself at the end of the twentieth century.[82]

Like White, Carl Amery contended that Christianity had been much more successful in changing the world than any other force in history, both through the actions of the faithful themselves and through its impact on dissidents and opponents. Even heretics and enemies of the Christian faith, Amery claimed, had in fact adopted and internalized some of the guiding ideas from the Judeo-Christian tradition. 'In the subjection of alien cultures, in the imposition of its own forms of thought and action, in the domination of nature', Christianity had clearly emerged triumphant. But the very success of Christianity threatened to become the undoing of the world, he argued, because the resulting pressure of humans on the natural environment had grown to such an extent, that – if the dire predictions of the Club of Rome were true – the future of the planet itself was now at stake.[83]

Next to this radical line, which painted the relationship between Christian values and technological advancement in rather crude, global terms,

---

[81] Forbes, 'Power', 606, Sambursky, *Physical world of the Greeks*, 244–245, Van der Pot, *Bewertung*, 36–41.

[82] White, 'Historical roots', 1203–1207.

[83] Amery, *Das Ende der Vorsehung*, 9–15.

there arose a subtler, more balanced approach. One of the first scholars to formulate a sophisticated thesis on the Christian foundations of technological advance was Ernest Benz in the early 1960s. Benz took a multilayered approach. He suggested that 'modern' technological advance in 'the West', which started in the period of the Renaissance, the Reformation and the Baroque, was built on a combination of presuppositions, all resting on Biblical sources, which were inherent to the Christian faith. These presuppositions were, in his view, the specifically Christian foundations of Western technology. First of all, Christianity conceived God as a Creator. Christians typically imagined God as a kind of craftsman: as a potter, architect or, later, watchmaker. Second, this notion of a Creator-God was associated with a particular appraisal of the natural world. The natural world merely ranked as a transient phenomenon without immanent god-like entities, which could be approached without timidity or fear. Third, man was assigned a special place in the Christian universe. Being made in God's image, it was assumed that he also had creative abilities. Fourth, as St Paul explained, man was even called to be a co-operator of God. Next, this conception of Man as a co-operator of God led to the idea that humans were bound to assist God in the preservation and domination of the world and the completion of His creation. According to Benz, it was this idea of man's duty that has been used time and again, since the Renaissance, to legitimize the pursuit of technological improvement. The Christian notion of the linearity of time and the finiteness of mankind's earthly existence, finally, gave the realization of this duty a particular urgency.[84] *Tempus fugit* trumped *carpe diem*.

The other major proponent of a more moderate interpretation of the role of Christianity in technological advances was, perhaps surprisingly, Lynn White. White did not always take a harsh view of the influence of Christian values and attitudes. While he never wavered from the belief that Christianity provided fertile soil for the growth of technology, neatly encapsulated in the title of his collection of essays published in 1968, *Machina ex deo*,[85] he did not invariably cast its impact on nature in terms of coercion, aggression or ruthless exploitation. In a more refined version of his theory, developed in the 1970s and 1980s and most succinctly expounded in his contribution to the *Dictionary of the Middle Ages* in 1988,

---

[84] Benz, 'I fondamenti cristiani della technica occidentale', 241–249, *idem*, *Christliche Endzeiterwartung*, 135–141; the German and Italian versions of Benz's thesis slightly differ on a number of points. I have followed the most elaborate version.

[85] White, *Machina ex deo*.

White argued that Christian views provided a favourable environment for technological innovation in three fundamental ways. These are the essence of the 'religious' explanation of the European 'invention of invention' to which David Landes referred.

Following Judaic tradition, Lynn White claimed, Christians held a conception of a Creator-God who was outside His creation and who had given mankind the right to use the natural world for its own ends. Moreover, Christians thought that time, also being a finite part of God's creation, was 'a unique resource to be conserved and utilized to man's advantage'. The more economically this resource was used – for instance, by increasing output in a given unit of time – the better God would be served. Finally, Christians, like Jews, conferred a spiritual value on manual labour. Manual labour was therefore seen as a virtuous and dignified activity, rather than as a lowly pursuit, as had been the case in Greco-Roman times. According to White, this revaluation of manual labour, which first took root in monastic circles in the Early Middle Ages and spread to wider parts of society in the thirteenth century, was 'integral' to the 'massive technological development' of the West.[86] To understand why Western Europe in the Middle Ages was much more technologically dynamic than the Byzantine world, which, in White's view, Benz's theory had signally failed to explain, it was also vitally important to take the 'distinctive qualities and moods that differentiate Occidental from Byzantine Christian piety' into account. The main differences were that monks in the West remained more deeply involved in 'secular affairs than in the East' and that Latin Christendom as a whole showed a more 'activist, voluntarist attitude' than Eastern Christendom, which attached higher value to 'right thinking' and contemplation.[87]

When Western scholars rediscovered the full scale of past Chinese achievements in technology, the issue of the religious context of technological development also became a matter of debate in Chinese historiography. In contrast with discussions on technological change in Latin Christendom, however, the debate in the Chinese case has concentrated at least as much on the phase of 'stagnation', supposed to have started

---

[86] White, 'Western technology', 662–663.

[87] *Idem*, 'Western technology', 663–664, *idem*, 'Cultural climates and technological advances in the Middle Ages', in: *idem*, *Medieval religion and technology*, 217–253, esp. 237–238, 241–252.

somewhere between 1300 and 1500, as on the phase of 'advance', which is assumed to have reached its peak between c.900 and 1300.[88]

The foremost expert on Chinese science and technology, Joseph Needham, suggested that especially Daoism, and to a lesser extent Buddhism, offered a highly favourable environment for the development of particular techniques, notably in chemistry, medicine and dietetics. Confucianism, in contrast, showed a much more reserved attitude towards technological innovation; and it was Confucianism, reshaped as 'Neo-Confucianism' under the Southern Song, that eventually became the dominant value-system in the Chinese State.[89] David Gellner, following a distinction made by Max Weber between 'puritanische Rationalismus' (focused on rational mastery of the world) and 'konfuzianistische Rationalismus' (focused on rational adjustment to a given state of affairs), suggested that the absence of a transcendentally-based 'Protestant ethic' may very well explain why China remained caught in Elvin's 'high-level equilibrium trap': an endogenous capitalist spirit did not emerge.[90] Nathan Sivin, however, has taken Needham to task for using the label 'Daoist' in a vague, almost meaningless way, and for failing to demonstrate a clear-cut connection between Daoism and a special interest in science and technology in any particular case.[91] Jean Pierre Voiret thought it doubtful whether Confucianism, Daoism and Buddhism were really very different with respect to their views concerning technology. Each tradition appeared to harbour both 'friendly' and 'hostile' ideas and attitudes vis-à-vis technology.[92] Mark Elvin disputed that 'late-traditional Chinese values and ideas' (especially Confucianism) were 'the key inhibiting constraints' for the continued development of invention, innovation and growth. In his view, China's 'failure' to create an endogenous form of industrial capitalism can be adequately explained by economic and ecological factors.[93]

---

[88] See the classic study by Elvin, *Pattern of the Chinese past*, parts two and three.

[89] Needham, *Science and civilization*, vol. II, 33, 131, 76–77, 496, 504–505, Needham, *Science in traditional China*, 15, 25, 29; for Needham's relative lack of sympathy for Buddhism, see Finlay, 'China, the West', 276 note 47, for interpretations of the relations between Confucianism and capitalism, see Brook, 'Weber, Mencius', Nelson, 'Sciences, civilisations'.

[90] Weber, 'Die Wirtschafsethik der Weltreligionen', 531, Gellner, 'Max Weber', 31.

[91] Sivin, 'On the word "Taoist" ', Cullen, 'Science/technology', 295–296.

[92] Voiret, 'Religionen und Technik in China', 88–90.

[93] Elvin, 'Why China failed to create an endogenous industrial capitalism', esp. 379, 389 note 3, *passim*, cf. Elvin, *Pattern of the Chinese past*, esp. 312–315.

Evaluation

The various theses discussed so far make the common claim that differences in technological performance between Europe and 'the rest', and within Europe itself, may at least to some extent be explained by religious factors. At this point, I should make my own position in this debate clear. Given the undeniable importance of religion in people's daily lives and the development of societies in general before the combined effects of the Enlightenment, political revolutions and industrialization in the nineteenth and twentieth centuries made themselves felt, I think that it would be truly astounding if religious traditions were found not have mattered at all in the history of technology before 1800. Those historians, theologians, philosophers, economists and other social scientists who suspect that religion had some influence on the development of technology surely have the benefit of plausibility on their side. The role of religious factors can only ignored at the risk of reducing our ability to understand technological change. And if we assume that technological change has indeed been a powerful force in the history of human societies (whether for good or for ill), then it is clearly highly relevant to study the impact of the religious context on technological change as well.

In the practice of historical research, however, it has proved to be very difficult to determine exactly how and to what extent religious factors actually affected technological development. In 1990, William B. Jones and A. Warren Matthews published an essay aimed at developing a classificatory scheme of relationships between religion and technological innovation. In this essay, they showed that there was hardly a possible connection between the two variables that had not been noted or studied in the scholarly literature published since the 1920s. Whether one thinks of 'disapproval or opposition to alien or novel technologies by a religion or religious leader', of 'approbation or promotion of inventions' by religious traditions in societies 'in which the religion is established', of the 'indirect influences' of religion by the 'reinforcement or promotion of conditions that favour (or hinder) an invention or innovations generally' or of 'detrimental effects of technological engendered alternatives upon religious institutions, beliefs and values' – the existing body of literature could always provide some ammunition to prove the point.[94] But Jones and Matthews also noted that some relationships had received more

94 Jones and Matthews, 'Toward a taxonomy', 3–23.

attention than others and that the 'indirect' or 'unintended' influence of religion upon technology, such as that suggested by White, Amery, Weber or Thorner, was harder to identify or to confirm than 'more-or-less direct effects'.[95] It was not easy to demonstrate conclusively that a religious tradition had indeed reinforced or promoted conditions that favoured, or hampered, the advance of technology.

In Chapter 1 of this book, I will argue that this failure has to do with the ways in which the issue is normally addressed. To date, the ways in which the impact of religious context on technological change has been approached have been singularly unfruitful. Studies on the impact of religion on technological innovation that start from the side of religion have tended to get stuck in casuistic discussions on the interpretations of texts or images and have frequently been in want of a rigorous comparative perspective. Studies starting from the side of technological and/or economic change, by contrast, are normally stronger on comparisons but have laboured under an excessive preference for calculating correlations – often based on rather crude sets of data – over carrying out stringent causal analyses; in these studies, religion is often treated as a kind of 'black box' whose contents need no further inspection.

Sorokin and Thorner, for instance, based their claims about the truth or falsity of Weber's thesis regarding the area of technological innovation on data borrowed from sources such as the *Handbuch zur Geschichte der Naturwissenschaften und Technik*, Alphonse de Candolle's *Histoire des sciences et des savants depuis deux siècles* and estimates of the populations of different European countries published by R.R. Kuczynski and Abbott Payton Usher, which are far too coarse to allow any meaningful conclusions. More recently, social scientists have studied empirical relations between religious variations (in the original Weberian sense or some adapted definition) on the one hand, and differences in rates of economic growth or timing of capitalist development between countries or societies in various parts of the world on the other. Examples of this approach include quantitative surveys by Barro and McCleary, Delacroix and Nielsen, Cavalcanti, Parente and Zhao, Becker and Woessmann and Blum and Dudley.[96]

---

95 Jones and Matthews, 'Toward a taxonomy', 14–17.

96 Barro and McCleary, 'Religion and economic growth across countries', Delacroix and Nielsen, 'The beloved myth', Blum and Dudley, 'Religion and economic growth', Cavalcanti, Parente and Zhao, 'Religion in macroeconomics'. An even more global comparative approach is taken in Elvin, 'Why China failed to create an endogenous capitalism'; his article does not address the Weber thesis as such, however, but rather the merits of Weber's methodological strategy in general.

The main shortcoming of this macroeconomic approach is that it involves the introduction of definitions and (often heroic) assumptions that reduce the chronological scope of the inquiry or treat religious factors as static, one-dimensional entities. Barro and McCleary, for instance, conducted a cross-country study on the relation between religion and economic growth, which operationalized religion as church attendance and belief in heaven and hell and which was restricted, by and large, to the period between 1980 and 2000. Delacroix and Nielsen tested Weber's thesis by examining the strength of the association between the religious composition of populations (defined as the percentages of Protestants and Catholics) and a wide range of dimensions of industrial capitalism in a number of European countries in the nineteenth century. Cavalcanti, Parente and Zhao examined the extent to which differences between Catholics and Protestants could explain delays in the beginning of industrialization by comparing the estimated performance of 'Catholic' and 'Protestant' economies (defined as a difference in belief as to whether prosperity during one's life is a barrier to reaching heaven or not), calibrated by a counterfactual study exploring what might have happened had England remained a 'Catholic' country after 1645, instead of becoming a 'Protestant' one. Becker and Woessmann argued that there *was* a relationship between Protestantism and economic success, but that this relationship was different from the one that Weber and many of his latter-day followers had imagined. Economic prosperity was indeed associated with Protestantism, not as a consequence of a particular work ethic, but as a result of the growth of human capital through better education, which was promoted by the spread of the Reformation. Differences in economic performance between Protestant and Catholic areas were related to variations in levels of literacy, not to variations in work ethic.[97] Berger and Woessmann based their argument on an almost static analysis of county-level of data from late nineteenth-century Prussia, without paying much attention to developments between the sixteenth century and 1871, to the history of literacy in Catholic areas outside Prussia since the sixteenth century, or to the history of literacy before the coming of the Reformation.

In contrast to these highly stylized exercises which barely take account of historical evolution, having studied a very long period (the end of the Middle Ages until the middle of the eighteenth century), Blum and Dudley asserted that they had discovered a kind of path-dependent effect

[97] Becker and Woessmann, 'Was Weber wrong'.

of religious beliefs on economic growth. They claimed to have found that 'a small change in the subjective costs of cooperating with strangers', due to a difference in religious belief, resulted in a 'profound transformation in trading networks', which benefited economic growth. The claim hinged on the assumption that 'Protestant' cities could be distinguished from 'Catholic' cities, that differences in growth between those groups of cities reflected disparities in economic growth and that these differences were related to a transformation in trading networks, which was in turn linked to the (supposed) fact that ascetic Protestants, believing in predestination, were less prone to default in contractual relationships with strangers than Catholics, who could easily obtain a pardon for any breach of contract through the sacrament of penance. Whether Catholics actually committed breaches of contract more often than Protestants, however, the study did not inquire.

In general, to my mind, approaches to this subject from various angles have suffered badly from a lack of interdisciplinary awareness. Scholars often seem barely acquainted with the state of research on the same topic in a different discipline. The debates on Weber's, White's or Benz's theses that are conducted in various disciplines are only interlinked to a limited extent. Although economists and other social scientists do take note of relevant work by historians, philosophers or theologians to some degree, and vice versa, discussions on these issues by and large evolve along separate trajectories.[98] Scholars working in this area seem to behave like Longfellow's 'ships that pass in the night, and speak to each other in passing, only a signal shown, and a distant voice in the darkness'.

[98] For example: Cook, *Matters of exchange*, discusses the validity of the Weber thesis without referring to studies by sociologists or economists, while the volume by Lehmann and Roth (eds.), *Weber's Protestant Ethic*, brings together essays by sociologists and historians, but not economists, and studies on the Weber thesis by economists cited above rarely refer to work by historians. Although sociologist Barbalet, *Weber, passion, and profit*, is more sensitive to changes in historical understanding and firmly attempts to place Weber's thesis in its historical context, he hardly refers to any recent work by historians. Sociologist Huff accepts the argument by economists Becker and Woessmann described above, but does not consider studies by historians on the same issue, see *Intellectual curiosity*, 309–310.

A Long-Term Comparative Approach

Aims, Concepts and Approach

This book aims to make an innovative contribution to current debates on the origins of the 'Great Divergence' between Europe and China and the 'Little Divergence' within Europe by analysing the relationship between the evolution of technical knowledge and religious contexts.[99] The key question is, to what extent can disparities in the evolution of technical knowledge between Europe and China and between regions within Europe between about 700 and 1800 be explained by differences in religious context? This book thus takes a long-term and comparative perspective. Its subject lies at the intersection of global history, the history of religion and the history of technology. The analysis concentrates on religion as a possible explanatory variable; the consequences of technological change for religion will not be considered.

Technical knowledge and religion are both complex phenomena, and a proper definition has to take this complexity into account. The concept of 'technology' or 'technical knowledge' used in this book refers to the abilities of people to control or transform nature for productive ends.[100] While this study will not be concerned with domestic appliances, nor with the skills that relate to the manipulation of money or people (such as financial techniques, military tactics or the practice of administration), the concept does include artefacts, drawings and any other aids that could be used in the control or transformation of nature for productive ends.

Religion, too, is a multi-faceted phenomenon, which finds expression in people's beliefs, practices and experiences, in institutions and in media, as well as in rituals, images and doctrines. The sociologist of religion Steve Bruce offers the following definition that, in his view, 'fits with broad contemporary common-sense reflection on the matter': 'Religion ... consists of beliefs, actions, and institutions, which assume the existence of supernatural entities with powers of action, or impersonal powers or processes possessed of moral purpose'.[101] As Bruce's critic, Charles Taylor, concurs, this definition forms a useful starting point for inquiry, provided one keeps

---

[99] The debate on the 'Little Divergence' in Europe does not yet have a counterpart in China, as I explained above.

[100] Davids, *Rise and decline*, 23; cf. Mokyr, *Lever of riches*, vii, Storper and Walker, *Capitalist imperative*, 50–51.

[101] Bruce, *Religion*, 7.

in mind that the line of demarcation between the natural and the supernatural is drawn less sharply in non-Western cultures and can change in the course of time.[102]

A study that seeks to advance our understanding of the relation between the evolution of technical knowledge and religious contexts thus has to cover a variety of aspects. It is impossible, however, to pay attention to all of the aspects mentioned in the definition in equal measure; religion, especially popular religion, is much too complex and effusive for that. Instead, this book focuses on four aspects which together cover a large part of the spectrum of possible interfaces between technical knowledge and religion, namely: visions on the uses of nature, human capital formation, the circulation of technical knowledge and technical innovation. Starting with the domain of ideas and values, it thus moves via the spheres of institutions, communications and movements of people to the field of practices and experiences where the actual technical innovation occurs.[103] For all of these aspects, we will look at the extent to which disparities in the evolution of technical knowledge can be explained by differences in religious context.

Changes in these aspects do not necessarily take place simultaneously and they can occur in different places at different times. To get a more exact idea of their timing and to discover how and to what extent changes in the various aspects built upon each other (for instance, technical innovation due to the circulation of technical knowledge, or the circulation of technical knowledge due to the formation of human capital), we have to make comparisons over time. The period covered should be long enough to allow us to observe both continuities and changes in technical knowledge and religious contexts in China and Europe before the onset of the Industrial Revolution at the end of the eighteenth century. From that perspective, the best point to begin our inquiry is the eighth century, when China's economic and technological achievement reached its first peak under the Tang dynasty (c.618–907) and Europe had started its long-term process of continuing innovation (whether drawing on indigenous sources or on a global common pool of knowledge). The study as a whole therefore covers the period from c.700 to 1800.

---

[102] Taylor, *A secular age*, 429, Bartlett, *The natural and the supernatural*.

[103] In a way, it covers religious aspects of what Patrick O' Brien has called 'regimes' for the discovery, development and diffusion of useful knowledge. O'Brien, 'Needham question updated', 10.

In spatial terms, this study takes a regional focus. From Chapter 2 onwards, the analysis in this book concentrates on a comparison of regions within Europe and China, rather than on a comparison of entire nations or continents. Units of the latter type would be either anachronistic or extremely hard to define. Within Europe and China, I have selected a group of regions which together show a gamut of changes in technical knowledge (including both accelerations and decelerations in innovation) and a diversity of religious contexts, notably including Catholicism and varieties of Protestantism on the European side[104] and Confucianism, Daoism, Buddhism and popular cults on the Chinese side.[105] Moreover, these regions are relatively well-studied and well-documented in the secondary literature and/or printed sources. In Europe, the set consists of North Italy, France, Spain, the Low Countries, England and Central Europe (roughly comprising the territories of the erstwhile Holy Roman Empire).[106] In China, the group comprises four regions, embracing parts of several provinces that in scholarship on Imperial China are commonly known as 'macro-regions': namely, the northern region incorporating parts of Hebei, Shandong and Henan, the area of the Lower Yangzi (also known as Jiangnan), the area of the Middle Yangzi centred on Hankou, and the lower east-coast area stretching from southern Zhejiang via Fujian to eastern Guangdong.[107]

I have not attempted to cover the whole range of technical knowledge in China and Europe or to compare entire 'material cultures', 'sociotechnological systems', 'technological cultures' or 'techno-scientific cultures'.[108] Some areas of technical achievement, such as transport or military technologies, will only be looked at in passing. Still, the spectrum of knowledge in China and Europe discussed in this book is quite broad and varied. Areas of technological activity covered include agriculture, water

---

[104] The comparison on the European side thus does not include Judaism, Islam or the Orthodox Church e.g.

[105] The comparison on the Chinese side thus does not include Islam, Nestorianism or Roman Catholicism e.g. On discussions about the religious dimensions of Confucianism, see Taylor, *Religious dimensions*, esp. 1–6.

[106] Most of these regions have also been dealt with in a comparative, collective volume that I co-edited with Jan Lucassen a few years ago, *A miracle mirrored*.

[107] Skinner (ed.), *City in Late Imperial China*, Skinner, 'Presidential address', Hartwell, 'Transformations of China', Spence, *Search for modern China*, 91–93, Ter Haar, 'Buddhist-inspired options', 95, Pomeranz, *Great Divergence*, 226.

[108] Cf. Lemonnier (ed.), *Technological choices*, Bijker, Hughes and Pinch (eds.), *Social construction*, Mukerji, *Impossible engineering*, esp. 206.

management, building and textile production, as well as metalworking, mining, machine making and book production.

Plan and Argument of This Book

The four chapters that follow form the core of this book. Each of these chapters is devoted to one of the aspects of the relation between technical knowledge and religious contexts introduced above: visions on the uses of nature, human capital formation, circulation of technical knowledge and technical innovation. The conclusion considers the implications of this study for current debates on divergences, and suggests a number of relevant issues and promising directions for future research.

The first chapter compares visions on the uses of nature. Taking the 'radical' and 'moderate' theses on the relation between Christianity and technology as a starting point, this chapter will analyse the extent to which religious traditions before 1800 showed differences or similarities in ideas and values concerning the uses of nature and, more importantly, the extent to which differences between religious traditions in these respects, if any, actually mattered for how technology was used in practice. Did religious traditions really differ in their visions on the uses of nature? And did people really deal with nature in a different way because the religious traditions to which they subscribed nourished different views on the uses of nature? I will suggest some strong reasons for doubting this. This angle of 'visions' and 'values' leads, to my mind, to a dead end.

In the next three chapters I will argue that differences in religious contexts nonetheless *did* matter for technological change, but that: (1) this effect was much more important for the *Great* Divergence than for the *Little* Divergence, and (2) differences in religious contexts primarily affected the evolution of technology by influencing institutions and underlying socio-political structures for the formation of human capital and the circulation and creation of technical knowledge. Variations in religious contexts, I will argue, were relevant to the emergence of differences in technical change and thus contributed to the Great Divergence. The chief point of impact, however, did not lie in the sphere of ideas, attitudes and values, but at the level of institutions, patterns of communications and movements of people connected with religious traditions. The three chapters that follow will deal, successively, with the relation between religion and the formation of knowledge and skills, the circulation of knowledge and technical innovation. In these respects, religion in the period up to 1800 was a powerful factor in technological change.

Figure 1. Illustration of a water-powered pounding mill in *Tiangong kaiwu* (The exploitation of the works of nature) by Song Yingxing, published in 1637.

Figure 2. Illustration of a foot-operated pounding mill in *Tiangong kaiwu* (The exploitation of the works of nature) by Song Yingxing, published in 1637.

CHAPTER ONE

RELIGION AND VISIONS ON THE USES OF NATURE IN CHINA AND EUROPE

This chapter takes the debates on the relation between technical knowledge and religious contexts outlined in the introduction as its starting point. It concentrates on the main bone of contention in these debates, namely the supposed impact of religious ideas and values on the development of technology. A number of scholars have sought the origins of the European 'invention of invention' in a particular set of religious values, specifically the 'Judeo-Christian subordination of nature to man' and related views on manual labour and the sense of time.[1] In this chapter, I want to analyse the extent to which religious traditions before 1800 showed differences or similarities in visions concerning the uses of nature and, more importantly, the extent to which differences between religious traditions, if any, actually mattered for the ways in which technology was used in practice. Did religious traditions really differ in their visions on the uses of nature? And did people really treat nature in a different way because the religious traditions to which they subscribed nourished different views on the uses of nature? I am sceptical about both these claims and in this chapter, I will argue as to why they should be rejected.

The debate on the differences between Latin Christendom and 'the Rest' will be looked at from three perspectives. Section one concentrates on the discussion on *internal* aspects of Latin Christendom and, in particular, on the different views put forward concerning the meaning of the Bible. Section two discusses literature on ideas and values regarding the uses of nature in the Mediterranean world that were current *before* the rise of Latin Christendom. In the third section, the focus shifts to the world *beyond* Latin Christendom. This section reviews the debate on the ways in which religious traditions other than Latin Christendom held different views on the uses of nature, and the extent to which these differences, if any, really mattered in practice.

---

[1] As noted by Landes, *Wealth and poverty*, 58–59; see also Van der Pot, *Bewertung*, 36–55, Popplow, *Technik im Mittelalter*, 105–110.

After reviewing the current debates about the presumed differences between Latin Christendom and other religious traditions, the next part of the chapter addresses the discussion on variations within Christianity itself. The central issues here are the supposed differences between the visions of nature of 'Puritans or 'Calvinists' and Catholics and the actual consequences of these differences for the ways in which nature was treated. The chapter concludes with a brief analysis of the underlying assumptions of these debates and sums up the implications of the argument for the following parts of the book.

Visions of Nature and the Great Divergence

Visions in Latin Christendom

Both the radical and the moderate versions of the theory on the relationship between Christian values and technological innovation have generated lively debate. In 1993, Elspeth Whitney reported that White's thesis in the *Science* article alone had been a subject of discussion in 'over two hundred books and articles by historians, environmentalists and philosophers'.[2] By the first decade of twenty-first century, 'ecotheology', studying the relation between religion and nature through time, had grown into a fully-fledged field with its own journals, conferences, handbook and professional organization.[3]

Much of the debate in ecotheology has revolved around the interpretation of the Bible that lay at the root of White's and Amery's arguments about Christianity's 'guilt' for the present ecological crisis. Critics contended that their representation of Christian views was in fact rather one-sided. They argued that it rested almost entirely on a very selective reading of the opening chapters of Genesis. The critics asserted that the Bible was much more ambiguous on man's relation with nature than White and Amery suggested. Dominion of nature could not simply be equated to exploitation. Even if Scripture taught that man was placed higher than the rest of creation, this did by no means imply that he could dominate the natural world at will. In a reader's letter to *Science* in response to the

---

[2] Whitney, 'Lynn White, ecotheology and history', 158, Stoll, *Protestantism*, 3–6.

[3] See e.g. the recent founding of the International Society for the Study of Religion, Nature, and Culture, the appearance of the *Journal for the Study of Religion, Nature, and Culture* and publication of *The Oxford handbook of religion and ecology*, edited by Robert Gottlieb.

appearance of White's provocative article, Ernest Feenstra put his finger on a crucial point: the Bible also related God's mandate to man to behave as a 'responsible steward' of the world of which he had been put in charge by the Lord. Humans were not free to loot, spoil or harass as they pleased; they were obliged to deal carefully with the natural resources that they had been given in trust.[4] Many adherents of the ecotheological movement have since laboured repeatedly on this matter of 'stewardship'.[5]

Outside the ecotheological fold, the issue of Biblical interpretation was given a significant twist by a number of other scholars, including Jacques Ellul, John Passmore, Robin Attfield and Katherine Temple. According to Jacques Ellul, a comprehensive reading of the book of Genesis (and not only of the story of Creation) showed that God did not grant humans more ample powers to deal with the natural world until *after* the Fall. Ruthless exploitation of nature was not inscribed in the pristine covenant between God and man. Technology, which allowed humans to dominate nature, was not an original attribute of man either, but just a temporary expedient that was only required after man had been expelled from the Garden of Eden.[6] The old Adam never felt a need for technology at all.

John Passmore thought that the 'critics of Western civilization' were right in claiming that 'there is a strong Western tradition that man is free to deal with nature as he pleases, since it exists only for his sake', but wrong in 'tracing this attitude back to Genesis', because Genesis, and the Old Testament in general, declared *both* that man is 'master of the earth' *and* that 'the world was good before man was created and that it exists to glorify God rather than to serve man'. It was not the Bible itself, Passmore argued, but the influence of Stoic philosophy, with its firm belief in the superiority of rational over irrational creatures, which ensured that the idea of the rightful exploitation of the natural world by humans became the dominant strand in Christian thinking. The notion of 'stewardship', meanwhile, did not spread in Christian circles before the seventeenth century and remained a minority tradition for a long time after.[7] However, Robin Attfield has claimed that this belief in 'stewardship' in fact

[4] Letter by Ernest Feenstra, *Science* 156 (12 May 1967), 737.

[5] Whitney, 'Lynn White, ecotheology and history', 159–161, Stoll, *Protestantism*, 3–6, 25–26.

[6] Ellul, 'Technique and the opening chapters of Genesis' and 'The relation of Man to Creation according to the Bible'. In their introductory essay 'Aspects of Christian exegesis', 29, Mitcham and Grote call Ellul's essays 'a withering critique of the interpretation of Christian revelation' by White.

[7] Passmore, *Man's responsibility for nature*, 14–18, 20, 27–32 and 40.

went back much further in time than Passmore assumed, namely to the Old Testament itself, and that it has been a more or less constant element in Christian thinking since Late Antiquity. Christian attitudes to nature in the late Roman and medieval periods, he concluded, were therefore much more diverse than White or Passmore allowed. The gentle strain in Christianity has never entirely disappeared.[8]

Katherine Temple, too, has disputed White's and Amery's views of the impact of the book of Genesis. Temple pointed out that the attitude of Christianity towards 'nature' was heavily influenced by Greek thinking, which first introduced the 'idea of nature' to the West, and that in the Jewish tradition, the opening chapters of Genesis were never read as a licence to exploit or plunder the natural world, or understood as a spur to develop new technologies.[9] The Bible could obviously be read in more than one way, and something else was needed to let one particular interpretation prevail. It should come as no surprise that Ausonius, author of the famed poem *Mosella* (on the River Moselle), which Mark Elvin hailed as a fine example of pre-Christian 'western nature poetry just before the transition to the Middle Ages', in which 'Nature itself [was] the main focus of interest', was not merely 'nominally' a Christian, as Elvin and others have claimed,[10] but a committed Christian. Christians in Late Antiquity could be enthusiasts of classical culture, too, as Alan Cameron has convincingly shown.[11]

The accuracy of the radicals' account of the changes during the Middle Ages and the adequacy of the explanation that they offered have also become the subject of debate. The claim that the spread of Christianity naturally implied the disappearance of animist beliefs and the dissolution of the view of the cosmos as a living, organic whole is no longer regarded as the truism that it was taken to be in the 1950s or 1960s. It is now widely assumed that many elements of these 'magical' ways of thinking were (probably deliberately) incorporated into Christianity in the Early Middle Ages and in fact survived until at least the sixteenth or seventeenth centuries.[12]

[8] Attfield, 'Christian attitudes to nature', 374–381, 385–386.

[9] Temple, 'Doubts concerning the religious origins'; a similar argument about the Jewish tradition has been made by Swetlitz, 'A Jewish commentary' and Landmann, 'Religiöse Bezüge zur Technik', 118.

[10] Elvin, 'Overview', 29–31.

[11] Cameron, *Last pagans*, esp. 404.

[12] Whitney, 'Lynn White, ecotheology and history', 166–167, Flint, *Rise of magic*, Thomas, *Religion*.

The moderates' arguments, expounded by scholars such as Ernest Benz and, once again, Lynn White, have been called into question, too. Doubt has been cast on the role played by Christianity and, in particular, the role of monasticism, in the supposed revaluation of manual labour during the Middle Ages. It is a matter of debate whether the status of work and the 'mechanical arts' in the Latin West really had changed by the end of the thirteenth century, and whether this change, in so far as it *did* occur, was causally connected to religious factors. Jacques Le Goff insisted that the Judeo-Christian tradition was no more free from ambiguities in its attitude to labour than other cultural traditions that left their mark on medieval culture. The Judeo-Christian heritage offered medieval people 'un arsenal idéologique contenant des armes à l'appui de toutes les positions, aussi bien en faveur du travail que du non-travail'. The fact that between the eleventh and thirteenth centuries, the scales were increasingly tipped in favour of the former can, in Le Goff's view, primarily be explained by changes in society at large, especially by the growth of cities and the increasing division of labour, rather than by some autonomous revolution in the religious sphere. The rise of new social groups led to a revision in the hierarchy of values.[13] Against Lynn White and Ernst Benz, George Ovitt argued that monastic communities in the Latin West always saw manual labour only as a means towards a spiritual end, and that the status of labour could only rise *after* it had been denuded of these spiritual connotations in the course of twelfth century. The Church's chief contribution to this revaluation was to adapt its theological views to the changing economic and social circumstances.[14]

But did the decisive change take place during the Middle Ages? The radicals' and moderates' accounts on the impact of medieval Christianity have also been undermined from another, perhaps somewhat unexpected, quarter. Feminist scholars such as Carolyn Merchant, Londa Schiebinger and others have argued that the crucial transformation in views on nature did not occur until *after* 1500. Throughout Antiquity and the Middle Ages, they insisted, nature was conceived as essentially female, but in two different, contrary ways.[15] Next to the image of nature as a wild, chaotic universe, which should be mastered and dominated, there existed the image of nature as a nurturing mother who provided for the

[13] Le Goff, *Pour un autre Moyen Âge*, 97, 115.
[14] Ovitt, *Restoration of perfection*, esp. 12–18 and chapters 2, 3, 4 and 5.
[15] Merchant, *The death of nature*, esp. chapter 1, Schiebinger, *The mind has no sex*, 122, Schiebinger, *Nature's body*, Keller, *Reflections*.

needs of humans in an ordered environment. Both these metaphors of domination and nurturing were not only to be found in Greek philosophy, but also in Renaissance culture. Each of these metaphors could serve as a 'controlling image' that operated as an ethical constraint or sanction on human behaviour. The image of the nurturing earth put moral restrictions on the exploitation of the natural environment; the image of domination, by contrast, sanctioned its relentless use. The latter image did not gain the upper hand until after – in Carolyn Merchant's words – the 'organically oriented mentality in which female principles played an important role' had been replaced by 'a mechanically oriented mentality that either eliminated or used female principles in an exploitative manner'. While the degree of exploitation – and the environmental degradation that went with it – in Antiquity and the Middle Ages gradually increased as a result of economic and technological changes, it was not until the sixteenth and seventeenth centuries that 'the tension between technological development in the world of action and the controlling organic images in the world of the mind' became so great that the 'female earth and virgin earth spirit' were finally 'subdued by the machine'.[16] If this feminist account is true, then the inference can be drawn (although this is not done by the feminist scholars themselves) that the ruthless exploitation of nature cannot be ascribed to the spread of Christianity as such. Commercialization, economic and geographical expansion and the mechanization of the world picture have been much more powerful factors in changing the uses of nature than the supposed impact of Latin Christendom.

The critique can be taken one step further. Neither White and Amery nor their opponents have looked at the actual extent of support for particular interpretations of the Bible in medieval and early modern Europe. While White's critics have not failed to uncover the evident bias in his account of Christian thinking on the natural world, by drawing attention to scriptural passages that betray a markedly different attitude, they have scarcely approached the issue from a historical point of view – even if White himself has countered textual criticism with a historical argument. After all, in response to Ernest Feenstra's remark (quoted above) that the Bible did not give a license for unlimited exploitation of the natural environment, but stressed man's duty to behave as a 'responsible steward' of the world, Lynn White insisted that the 'historical impact of Christianity upon ecology has depended not on what we, individually, at present, may

[16] Merchant, *The death of nature*, 2–4, 41.

think that Christianity should have been, but rather upon what the vast "orthodox" majority of people who called themselves Christians have in fact thought it was'.[17] But did they think in this way? Did most people in the Middle Ages or the early modern period really not care about the notion of 'stewardship' at all? That is a matter for historical inquiry. White has not produced any evidence for his claim, nor has anybody else, as far as I am aware. The fact is that at present, no one knows what the vast majority of Christians in the past were thinking about at this particular issue. Claims to the contrary are merely suppositions. Unfortunately, the debate has been silent on this crucial point.

The follow-up question has not figured in the discussions either; even if we knew for certain that most Christians were indeed thinking what Lynn White assumed they were thinking, we would still have to find out *why* these Christians adhered to this particular interpretation of Scripture at all. Ignoring this question would imply that one assumed that this interpretation of the Bible was somehow a natural choice; and it is entirely unclear on what grounds such a supposition would be warranted.

Visions before Latin Christendom

In contrast with the ecotheologians, who (as Whitney pointed out), despite all of their disagreements, shared the basic assumption made by White, Amery and Benz that 'religion was the prime determinator of human action',[18] a second group of critics of the theory concerning the relationship between Christian values and technological innovation denied that the rise of Christianity implied a notable change in attitudes towards technology. They insisted that ideas and practices that are identified with the spread of Latin Christendom had a much longer history than White, Amery and others assumed. This second strand of criticism had in a sense been foreshadowed in the seminal work of Clarence Glacken on the history of Western thinking about nature, *Traces on the Rhodian shore*,[19] which appeared almost simultaneously with the provocative essay by Lynn White. On the basis of a wide-ranging study of classical authors, Glacken showed that already 'in the ancient world, there was a lively interest in natural resources and how man could exploit them'. In the

---

[17] Rejoinder by Lynn White to letter by Ernest Feenstra, *Science* 156 (12 May 1967) 737–738.

[18] Whitney, 'Lynn White, ecotheology and history', 161.

[19] Glacken, *Traces on the Rhodian shore.*

Hellenistic period, in particular, 'a philosophy of resource development seem[ed] to go hand in hand with the economic and political aspirations' of princely rulers.[20] The feminist account presented by Carolyn Merchant and others perfectly concurs with this argument. Images of domination were not a creation of Latin Christendom; rather, they were rooted in Greek philosophy.

Moreover, in Antiquity, were technology and labour really held in lower esteem than during the Middle Ages? The 'mainstream view', as Serafino Cuomo has called it, namely 'the stubbornly common picture according to which ancient technology was marginal to society and economy, its practitioners were widely despised and its activities looked down upon',[21] has been seriously called into question by recent research. Cuomo has argued that the sense of menace from 'base technicians' that was felt by the proponents of an aristocratic ethos of society, such as Plato and Aristotle, points to the centrality, rather than the marginality, of technicians in Classical Athens. Even though the supporters of an aristocratic order might have despised them, Athens could not have functioned without technicians to build its ships, fight its wars and cure its people: 'without *technai* and consequently without technicians, there would be no human civilization as such' – something that people at the time were very well aware of.[22]

The idea that there was a fundamental difference in thinking about labour and technology between Antiquity and the Middle Ages has been undermined as well. After a meticulous analysis of various sorts of literary sources, ranging from Stoic writings from the time of the Principate to thirteenth-century sermons, Birgit van den Hoven concluded that there was, in fact, no radical change in thought on these subjects.[23] Thus the very distinction between a 'static', 'conservative' climate in Antiquity and an 'innovative', 'modernizing' mentality in the Middle Ages, which lay at the heart of White's work,[24] has been eroded. And with that, the causal link between the rise of Christianity and the acceleration of technological change has been thrown into doubt, too.

---

20 Glacken, *Traces on the Rhodian shore*, 118, 712.

21 Cuomo, *Technology and culture*, 4.

22 Cuomo, *Technology and culture*, 38–30.

23 Van den Hoven, *Work in Ancient and Medieval thought*, 254–255.

24 Hall, 'Lynn White's *Medieval Technology and Social Change* after thirty years', 92–94; cf. also Amouretti, 'L'attelage', 227–228, on the ideological background of the revaluation of the Middle Ages and the depreciation of Antiquity between the 1930s and 1980s.

Another ingrained notion about Antiquity, namely the idea that 'something "blocked" the ancient mind and prevented it from making those connections between technology and economy, or technology and wider applications',[25] has lost its credibility as well. Since the end of the 1970s, the view of the Greeks and Romans as technical practitioners has become more widely accepted. Nowadays, the Ancients are rarely castigated for their supposed lack of technological innovation. Far from being portrayed as inveterate conservatives accustomed to relying on slaves rather than machines to do their work, they are now increasingly esteemed for their achievements in many different fields of technology, including some of the very fields which White singled out in *Medieval technology and social change* as prime examples of medieval creativity, namely the use of water-powered mills and the adoption of a better form of horse-traction.[26] A vast amount of archaeological evidence is now at hand to prove that the breakthrough of the watermill actually occurred long before the Early Middle Ages, contrary to what White, following Marc Bloch, mistakenly believed. Watermills were already quite widespread in the Roman Empire in the first century AD, and they were also used for other purposes than grinding corn, such as sawing marble. The cam, a key element for industrial applications of waterpower, was certainly known in Hellenistic times.[27] Indeed, Egyptians, Assyrians, Greeks and Romans appear to have been much more flexible and creative in devising effective solutions to the problem of harnessing horses for traction than White, basing his own account on what was for long the authoritative work by Lefèbvre des Noëttes, gave them credit for.[28]

Moreover, products and processes whose introduction or spread used to be regarded as achievements of the 'Islamic' era have turned out, in fact, to have their roots in the pre-Islamic period, i.e. in Roman or Sassanid (Persian) times. Muslims did not construct the irrigation system in southern Spain from scratch, but built on a pre-existing Roman network. The diffusion of crops like durum wheat, rice, sorghum and artichoke,

---

25 Cuomo, *Technology and culture*, 3–4.

26 White, *Greek and Roman technology*, 172–174, Greene, 'Technology and innovation', 22–33, Greene, 'Technological innovation', Lewis, 'The Hellenistic period', Wikander, 'The Roman empire', Watson, 'Machines', 6–24.

27 Wikander, 'Water-mill' and 'Industrial applications of water-power', Holt, 'Medieval technology and the historians'. Cf. the classic article by Marc Bloch: 'Avènement et conquêtes du moulin à eau'.

28 Amouretti, 'L'attelage dans l'Antiquité'; the work referred to is Lefèbvre des Noëttes, *L'attelage, le cheval de selle à travers les âges*.

once hailed as evidence of an Islamic Agricultural Revolution, was already well under way in the Mediterranean world in Antiquity – long before Islam appeared on the scene.[29] Lynn White's thesis that the crucial transformation in technology in the Western world occurred during the Middle Ages thus appears to be seriously flawed.

Visions in China

Both the radical and moderate versions of the theory on the relationship between Christian values and technological innovation assume that the decisive technological changes in history, and the environmental degradation that followed, were somehow dependent on the rise of Latin Christendom. It was supposedly the rise of Latin Christianity that generated the special attitudes towards man's relation to nature that eventually resulted in seminal technological breakthroughs and the ruthless exploitation of nature.

But was this indeed the case? One of the first critics of Lynn White's thesis on the Christian roots of our ecological crisis tersely mentioned the fact that human overkill leading to the destruction of a large part of the mammalian fauna in Africa and North America was already taking place, respectively, some 50,000 and 12,000 years ago, and suggested that this 'clearly demonstrated that the Judeo-Christian tradition [was] considerably older than generally assumed'.[30] This jest, which received no rejoinder from White, raised of course a very serious question, namely: whether the exploitation of nature by technological means really only started with the rise of Christianity, or whether it also occurred in times and places where other religious traditions reigned supreme. Yet, until recently, this question hardly received any consideration at all. The problem of technological innovation has largely been reduced to a problem relating to Latin Christianity, as if the rest of the world hardly counted.

The striking neglect of possible relationships between technological innovation and religious traditions other than Latin Christianity cannot entirely be ascribed to a sheer lack of familiarity with non-Christian religions. Ernest Benz, for example, was certainly no stranger to Buddhist thinking. According to his own account, his awareness of the supposed

[29] Decker, 'Plants and progress', esp. 190–191, 206, Wilson, 'Machines', 9. These recent findings challenge the influential theses put forward by Glick, *Irrigation and society in Medieval Valencia*, and Watson, *Agricultural innovation*.

[30] Letter by Cesare Emiliani and Shale Niskin, *Science* 156 (12 May 1967), 738.

role of Christianity in technological change was aroused by his observations as a visiting professor in Asian countries. Thanks to his experience in Asia, he developed a keener eye for the role of religious factors in the varied responses of Asian societies to 'Western' technology.[31] A more likely explanation for the almost automatic association between technological innovation and Christianity is the Euro-centric vision that, until lately, dominated historical studies in general. Given the persistent focus on the 'rise of the West' as the supposedly central theme of the past thousand years of world history, and given the fact that the West was overwhelmingly Christian, the temptation to seek a possible relation between technological innovation and religious context exclusively in special features of the Christian tradition must have been nearly irresistible. In his *Science* article of 1967, Lynn White assumed without more ado that 'both modern technology and modern science are distinctively *Occidental*' and Europe possessed a 'technological superiority' over the rest of the world by 1500.[32]

While most historians would still readily agree with the former part of the claim, this is no longer true for the latter. The works by Needham, Elvin, Hartwell, Bray and others on technological innovation in China under the Tang, Song and Ming dynasties, and by Glick, al-Hassan, Hill and others on technical change in Islamic states up to the sixteenth century, have destroyed the foundations of this confident assertion forever.[33] Nowadays, it would be preposterous to state that Europe had manifestly gained a technological lead over China or the Islamic world at the dawn of the early modern era.

Moreover, recent work on developments in China and Japan has demonstrated that Latin Christendom did not have a monopoly on the idea of manipulating natural forces for human ends, or on the practice of harsh exploitation of the natural environment. Robert Weller has observed that Mainland China and Taiwan (which can hardly be called Christian-dominated societies) did not see any signs of environmental consciousness, environmental protection and a sensitivity to the intrinsic qualities

---

[31] Ernest Benz, 'I fondamenti cristiani della technica occidentale', 241, 'Christliche Endzeiterwartung', 135.

[32] White, 'Historical roots', 1204.

[33] E.g. Needham, *Science and civilisation in China* (many volumes), Elvin, *Pattern of the Chinese past*, Hartwell, 'A cycle of economic change in Imperial China', *idem*, 'Demographic, political, and social transformations of China', Bray, *Technology and society in Ming China*, Glick, *Irrigation and society in Medieval Valencia*, al-Hassan and Hill, *Islamic technology*, Hill, *Studies*; see also Long, *Technology and society in the medieval centuries.*

of 'nature' before the middle of the 1980s.[34] Like Japan since the Meiji restoration,[35] both the People's Republic of China and Taiwan in the post-war era adhered to a purely utilitarian approach to the natural environment in the name of economic 'progress'. The 'supreme indifference to the destruction of nature by the process of industrialization', which was said to be a hallmark of the Meiji regime, also characterized the communist and nationalist governments in China after 1950. The spirit of the period is perhaps best captured by the slogans which the Maoist regime used to promote its crash programmes of building giant dams and creating heavy industry from scratch: 'Man Must Conquer Nature', and 'Battling with Nature is Boundless Joy'.[36] It was the interaction between Western ideas about the conservation of nature and Chinese traditions that led to a change in attitudes and policies and behaviour at the end of the twentieth century, including the rise of 'nature tourism' and the creation of national parks.[37]

The exploitative attitude to the natural environment that typified these East Asian societies during their rapid economic transformation in the late nineteenth and twentieth centuries, however, did not first arise in the period of industrialization. Similar ideas had also existed in earlier times. Visions of nature in Japan or China have never formed a uniform, static whole, infused by a spirit of oneness and harmony between human beings and the natural environment. Tessa Morris-Suzuki found in Tokugawa Japan a wide variety of concepts about the relation between humans and nature. To be sure, some Japanese thinkers in the eighteenth century expressed the kind of notions about unity between man and nature which 'eco-nationalist' historians and anthropologists have claimed as quintessentially Japanese. But the late Tokugawa period also saw the appearance of other visions, which emphasized the human duty to 'open up nature' (*kaibatsu*). According to this view, the use of natural resources for human ends should not be censured as a violation of a sacred bond, but should rather be seen in a positive light. The former outlook has been associated with Daoist, Buddhist and Shinto traditions, the latter with Neo-Confucian influences.[38]

---

34 Weller, *Discovering nature*, 2–12.
35 Morris-Suzuki, 'Concepts of nature', 82–83.
36 Morris-Suzuki, 'Concepts of nature', 96–97, Weller, *Discovering nature*, 49–53.
37 Weller, *Discovering nature*, 64–101.
38 Morris-Suzuki, 'Concepts of nature', 82–84, 88–89, 94–96.

Regarding China, Robert Weller observed that most visions of nature in China are 'perfectly comfortable with manipulating natural forces for human benefit', even though the forms taken by these visions could vary widely.[39] In this respect, Confucian, Daoist or Buddhist traditions differed less than it would first appear. Derek Bodde distinguished seven approaches to nature in the history of China, ranging from rejection via exploitation, analysis and animistic and moralistic views to a total, mystical union with the natural world. These approaches were neither mutually exclusive nor solely linked to particular religious traditions.[40] The idea that in Chinese thought, 'man' and 'nature' have always been viewed as a seamless whole and that harmony with nature has invariably been seen as the ultimate ideal, turns out to be largely a construction of twentieth-century scholars.[41] Views of nature in Chinese philosophy have, in fact, varied from the very beginning. Heiner Roetz has shown that the Confucian philosopher Xun Zi, who flourished in the third century BC, stressed the dichotomy, not the unity, between 'man' and 'nature' and vigorously argued for active human control over nature.[42]

'Nature' itself is not, of course, an unambiguous, culture-free concept. Before the end of the nineteenth century, there was not yet a generally accepted Chinese translation of the English word 'nature'. 'Nature' in the sense of 'an essential quality of something', 'the inherent force directing the world or humanity' or 'the material world itself' did not have an equivalent in China. The Chinese language *did* have words that covered *some* aspects of the English (or by extension, European) concept of nature, such as words denoting 'flow of energy' or 'existing thing not caused by something else', but it did not yet have a single, overarching term that captured its entire meaning.[43] The Chinese language had 'a wide range of expressions to refer to the observed or imagined regularities of the physical world of things'.[44] However, evidence for the existence of an idea of 'laws of nature' in China before the twentieth century is exceedingly rare.[45] The difficulty of translating between English and Chinese betrays the fact that

39 Weller, *Discovering nature*, 40.
40 Bodde, *Chinese thought*, 309–332.
41 Roetz, *Mensch und Natur*, part I, esp. 30–49.
42 Roetz, *Mensch und Natur*, part II, esp. 284–285, 347; see also Roetz, 'On nature and culture'.
43 Weller, *Discovering nature*, 20–23, Sivin and Lloyd, *The way and the word*, 200.
44 Harbsmeier, 'Towards a conceptual history', esp. 225 and 252.
45 Bodde, *Chinese thought*, 344.

views of nature in China were different from the concept that became predominant in Europe from the late seventeenth century onwards.[46]

Yet, the absence of a comprehensive Chinese term for 'nature' and the evidently multiple character of views of nature in China does not mean that there were not some remarkable similarities between the concepts that emerged in the European and Chinese contexts. Although Europe and China did not share a *positive* concept of nature, their ideas on a world of phenomena *beyond* the ordinary one showed a certain resemblance. For a long time in both China and in Latin Christendom, people assumed as a matter of course that there existed a 'supernatural' world next to the 'natural' one. This supernatural realm, peopled by godlike powers, immaterial beings and deceased humans, was distinct from, yet also interconnected with, the visible physical environment. Entities from this outer sphere could exert influence in the natural world, and humans could look forward to a more or less prolonged afterlife in the world beyond perception.[47] Both China since at least the tenth century and Latin Christendom since the twelfth century have known the concept of a 'temporary hell' or a 'purgatory' as a space where the dead would be punished for a while for sins committed during their lifetime.[48] This notion of punishment, in its turn, was in both cases connected with a concept of 'sin' as a violation or transgression of rules from a moral code that was held to be generally valid.[49]

Both in China and in Latin Christendom, moreover, all sorts of rituals developed to placate the powers from the supernatural world, reduce the penal burden for the dead or obtain favours or assistance for the living. These rituals were often, but not exclusively, performed by mediators such as priests, exorcists, fortune-tellers and other 'ritual masters'; and lay people could also try to contact or placate supernatural powers themselves by saying prayers, wearing charms or venerating images or statues. No single 'ritual master' or organization of 'ritual practitioners' enjoyed a monopoly on mediation: laymen could usually choose between the services of different practitioners. Under the Southern Song, religious specialists were 'moving from place to place, from marketplace to marketplace, offering their services to those in need'; they were asked to 'recite sutras,

46 Daston, 'The nature of nature', 157–168; Elvin, 'Overview', 15–16.

47 Daston, 'The nature of nature', Bartlett, *Natural and supernatural*, Davis, *Society and the supernatural*.

48 Teiser, 'Growth of purgatory', Eberhard, *Guilt and sin*, 17, 26–59.

49 Eberhard, *Guilt and sin*, 12–23, 60–75.

conduct ceremonies . . . write charms' and occasionally even 'to bring rain or to perform other miracles'.[50] This elaborate complex of demand and supply for rituals and mediators in China shows a striking resemblance to the world of 'religion' and 'magic' in medieval and early modern Europe famously described by Valerie Flint and Keith Thomas.[51] Both in China and in Europe, *literati* were concerned to keep the influence of this supernatural world on people's minds somehow in check by defining rules for the interpretation of otherworldly signs or phenomena, or by controlling the mechanisms or institutions of mediation.[52] 'Ritual masters' should not run amok, lest society or the state should be rocked by unrest or rebellion. The difference between China and Europe rested primarily in the way and extent to which the boundary of the 'supernatural world' was pushed back in favour of something called 'nature'.

Turning from ideas and attitudes to actual practices, the fact that immediately strikes one is that ruthless interventions in the natural environment were at least as frequent in China as in Latin Christendom. Nicholas Menzies reminds us that 'it is not easy to reconcile [the] apparent sensitivity to the natural world' displayed in Chinese philosophy, poetry or painting 'with the physical evidence of millennia of exploitation of the natural environment'. Even though on multiple occasions and in many different places, households or government officials, for example, *did* show genuine concern for the conservation, maintenance and careful management of forests, the stark reality was one of the progressive destruction of forests in large parts of China.[53] According to Mark Elvin, some environmental degradation already took place during the 'medieval economic revolution' between c.900 and 1300, when the empire saw an impressive spurt in agricultural productivity, urbanization and mechanization, and improvement in transportation and infrastructure. The destruction of forests, soils and wildlife habitats reached its nadir during the population explosion in the late-imperial period, especially in the most technically advanced regions of the country.[54] The pressure on the environment in Late Imperial China, Elvin estimates, was much higher than in Western Europe as a whole at the end of the eighteenth century, and perhaps even somewhat higher

50 On China: Davis, *Society and the supernatural*, 5–13, Hansen, *Changing gods*, 40–47, 75, 78–80.

51 Flint, *Rise of magic*, Thomas, *Religion and the decline of magic*.

52 Daston, 'The nature of nature', 167–171, Elman, *On their own terms*, 20–21.

53 Menzies, *Forest and land management*, 4, 21–33, 136; see also Roetz, *Mensch und Natur*, 80–85.

54 Elvin, *Retreat of the elephants*, 6–7, Elvin, *Pattern of the Chinese past*, part two.

than in its most advanced states, England and the Netherlands. 'Particular characteristically Chinese beliefs and perceptions' concerning the natural environment can in fact hardly be considered a relevant factor in determining how the Chinese actually used the resources of nature.[55]

Thus the roots of ecological crises in the present, or the past, cannot be exclusively sought in Latin Christendom. Exploitative attitudes towards the natural environment have also existed in other traditions than the Christian one, even though 'nature' itself was conceived in a different way from that which became normal in early modern Europe. Like in Europe, technological change in China brought environmental degradation in its wake. Neither technological innovation itself nor its deleterious effects for the natural environment can thus simply be ascribed to the typical traits of a supposedly widely shared vision of nature in Latin Christendom. In this respect, the radical and moderate versions of the theory on the relationship between Christian values and technological innovation can no longer be sustained.

Visions of Nature and the Little Divergence

Lynn White, Carl Amery and Ernest Benz were primarily concerned with the relationship between technological change and Latin Christianity as a whole. Variations *within* the Latin Christian world largely remained outside their range of vision. A number of scholars have suggested, however, that it was not Latin Christianity as such but rather a specific variety of Christianity, namely Puritanism or Calvinism, that created a favourable environment for technological innovation. This thesis implies that even if variations in visions of nature in particular religious traditions did not matter much in explaining variations in technological change *between* Europe and Asia, they could still be of importance in explaining variations *within* Europe. Differences in visions of nature between religions could have been instrumental in producing the Little Divergence rather than the Great Divergence.

Although this thesis perhaps has a Weberian tinge about it, it was not formulated by Max Weber himself. Weber's studies on religious sociology, one recalls, contained only the faintest of hints of a possible link between Puritanism and technological advancement. His most suggestive remark

---

[55] Elvin, *Retreat of the elephants*, 460, 470–471.

on the subject is the assertion that 'der puritanische Rationalismus', in contrast with 'der konfuzianistische Rationalismus', meant rational mastery of the world, not rational adjustment to a given state of affairs.[56] The implicit message of this remark – namely that Puritans, or Calvinists, were more bent on mastering the natural environment than Confucians or Catholics – was in recent times cast in a more explicit form by Mark Stoll and Benjamin Elman. Mark Stoll claimed that even if Protestants and Catholics largely shared the same views on man's dominion of nature, Protestantism – and especially Calvinism – actually stimulated the exploitation of nature more than Catholicism, because in its campaign against the 'non-Biblical magical or semi-magical practices' of Catholicism, it simultaneously eliminated many barriers that up until then had checked the full implementation of this exploitative drive. In the Calvinist world, the spell of the supernatural was broken earlier than in countries where Catholicism held sway.[57] In his book on science in China, Benjamin Elman suggested that Jesuits and Confucian elites in their visions of nature had more in common with each other than with Protestant elites in Europe:

> Among late imperial Chinese elites, and the Jesuits in China ... the otherworldly as a counterpart to the natural world never quite diminished as much as among secular Protestant elites in northern Europe.... The naturalization of unusual phenomena among Ming literati, as for the Catholic Jesuit missionaries who were tied to a new counter-Reformation vision of the natural world, remained incomplete.... Among Protestant men of science the boundary between the natural and the artificial dissolved into the notion of God as a great artisan, who could be mimicked by the great artist or man of science.[58]

But were denominational differences within Christianity in visions of nature really that great? To my mind, claims such as those of Stoll and Elman present an overly monolithic and static view of Catholicism in the early modern era. They are still based on the traditional assumption that Tridentine Catholicism formed a kind of unchanging, homogeneous, authoritarian entity, which was at best out of touch with, if not fundamentally hostile to, new developments in the world of science. That supposition can no longer be maintained in the light of recent research. It is hard to reconcile with evidence on regions or organizations which are often regarded as the very heart of Tridentine Catholicism, such as Italy,

56 Weber, 'Die Wirtschafsethik der Weltreligionen', 531.
57 Stoll, *Protestantism*, 30–31, 76.
58 Elman, *On their own terms*, 21.

the Spanish Empire or the Society of Jesus. And if core areas and actors do not fit the picture, what is left of the thesis as such?

The similarities in attitudes to the supernatural between Jesuits and Confucian elites that were observed by Elman were superficial, rather than fundamental. Qiong Zhang has shown that contrary to the traditional assumption, Jesuits in China were indeed as much concerned with exorcising demons and performing miracles as with demonstrating the benefits of 'Western learning' and striving for accommodation with indigenous culture. But their engagement with the supernatural primarily served a strategic purpose: showing that Christianity and scholarship in Latin Christendom was better equipped to deal with such otherworldly phenomena than the established religious cults in China. 'The Jesuits employed science as God's weapon against superstitions, which they defined as misguided (that is, unorthodox) beliefs', Zhang remarked.[59] The Rome-based Jesuit Athanasius Kircher, regarded as the greatest luminary of Europe in the Baroque age, waged a continuous battle against 'demonic magic'.[60]

Moreover, even supposing that Catholics were truly less prepared to relinquish belief in the supernatural or that they, more than Protestants, remained attached to magical or semi-practical practices, there is no evidence that this attitude actually mattered much for the way in which they approached the natural world. Continued receptivity to otherworldly phenomena was in reality not incompatible with an empirical approach to nature or with a desire to master nature for human ends. When Spanish scholars such as the *cronista de Indias* Gonzalo Fernández de Oviedo and the Jesuit José de Acosta in the sixteenth century composed the first general surveys of nature in the Americas, they displayed a fairly practical, open-minded attitude in their descriptions of the flora, fauna and natural occurrences in the New World and in their efforts to explain the underlying rhythms and patterns. They viewed nature as God's creation, but one that was accessible to human understanding. In their vision, studying nature was a means to praise and reach God. Personal experience and collective accumulation of knowledge, according to Oviedo, would ultimately result in an answer to every question about nature, if need be by the special grace of God himself. Acosta thought that experience fed information to human imagination and that imagination, guided by reason anchored in faith, would then order this information into universal knowledge. He

[59] Zhang, 'God, demons, and miracles', 4–6, 35–36.
[60] Findlen, *Possessing nature*, 81.

was convinced that nature, being organized 'as a machine' according to God's laws, could not be altered by man, but that inferior creatures in nature, such as plants or animals, were destined to serve higher ones, that is, humans.[61] Ulisse Aldrovandi, professor of natural history at the university of Bologna from 1561 till 1605, nicknamed 'the Bolognese Aristotle' and 'the Pliny of his time', did not only collect thousands of natural objects in his famous private museum as a basis for writing the definitive history of nature, but also performed dissections and conducted experiments to put established truths about nature to the test and, through sensory experience, improve man's understanding of the natural world.[62]

Such an ambition was too limited for Athanasius Kircher. At the Jesuit College in Rome, where he held the chair of mathematics from 1638 until 1680, Kircher created a museum that, in its day, became even more celebrated than Aldrovandi's *studio* in Bologna. Collecting naturalia, artefacts and manuscripts from all over the world formed the basis for a much grander project. The underlying purpose of this encyclopaedic effort was to decode the divine order of the cosmos, to achieve mastery of nature and thereby vindicate the Catholic Church's claims to supremacy. Kircher did not merely aim, like Aldrovandi, to understand and demystify nature, but he also aspired to replicate and manipulate natural phenomena. The machines and automata in the Roman College Museum, many of which had been designed by Kircher himself, were intended to prove the point. The presence of this showcase in the very heart of the Catholic world sent a powerful message. There was no institution on Earth more capable of 'possessing' nature than the Church of Rome.[63]

Like Dutch or English Protestants, Catholics in Italy or the Spanish Empire were often engaged in efforts to control the natural world for human ends. In the seventeenth and eighteenth centuries, for instance, scholars and engineers in Italy were busy developing new theoretical and practical approaches for the management of rivers. The 'theoretical turn' in hydraulics began in the Papal States in the 1620s and reached the Venetian Republic a few decades later.[64] The founding father of the 'science of waters' was a Benedictine monk who taught mathematics at Pisa and Rome, Benedetto Castelli.[65] Thanks to the combined efforts of a stellar

---

61 Barrera-Osorio, *Experiencing nature*, 106–120.
62 Findlen, *Possessing nature*, 23, 54, 62–63, 199–207, 211–213, 253, 279–280.
63 Findlen, *Possessing nature*, 46, 54, 78–93, 126–128, 350.
64 Maffioli, *Out of Galileo*, 419–20.
65 Maffioli, *Out of Galileo*, 41, 45–51.

group of other Italian scholars, including Evangelista Torricelli, Geminiano Montanari, Domenico Guglielmini, Guido Grandi, Bernardino Zendrini and Giovanni Poleni, in the following decades, this new approach to river hydraulics was expanded into an elaborate corpus of general concepts, principles and laws relating to the motion of water.[66]

With regard to Spain, recent research has amply refuted the traditional presumption that Spaniards in the sixteenth and seventeenth centuries were not interested in technology or science.[67] The Black Legend once again proved to be no more than a legend. True, many experts in mining, navigation, architecture, artillery or medicine recruited by the Spanish Crown in this period hailed from abroad.[68] While it is unlikely that all of them were Catholics, a substantial part of the innovative or routine operations in mining, natural history, agriculture or navigation was nevertheless carried out or supervised by Catholic Spaniards.[69] Acosta and Oviedo were not lonely figures; curiosity and expertise were not wanting. While in the 1570s and 1580s, the 'Bolognese Aristotle' Aldrovandi failed to get Philip II's support for an expedition to the New World, between 1570 and 1577, the Spanish court physician Francisco Hernández, by order of the King, carried out a comprehensive exploration of the flora and medical customs in the Caribbean and New Spain.[70]

The technological literature does not betray a significant difference in vision between Protestants and Catholics either. Marcus Popplow, analysing the rhetorical means with which technological endeavours were justified in 'theatres of machines' in the early modern period, concluded that for a long time, Biblical references hardly played a role at all. Before c.1620, when this literature was still heavily dominated by Italian authors, religious arguments were in fact seldom used. Mechanical inventions were mostly praised for their 'newness', 'usefulness' or 'ingenuity', not for their function in the restoration of Paradise on Earth. Religion did not become a prominent plank in the justification of technological projects until later in the seventeenth century.[71] This was not a purely Protestant manoeuvre, however. In his study of the uses of biblical references in 'theatres of

[66] Maffioli, *Out of Galileo*, 14, and Tables 1.1. and 1.2.

[67] Goodman, *Power and penury*, 262, 264, Navarro Brotóns and Eamon, 'Spain and the Scientific Revolution', 33.

[68] García Tapia, *Ingeniera y arquitectura*, 41.

[69] Goodman, *Power and penury*, 262–264, Barrera-Osorio, 65–80.

[70] Goodman, *Power and penury*, 234–236, Barrera-Osorio, *Experiencing nature*, 80, 121, Findlen, *Possessing nature*, 314–315.

[71] Popplow, *Neu, nützlich und erfindungsreich*, 3–5, 196.

machines' published in various European countries between c.1550 and 1750, Ansgar Stöcklein found that a tendency to base aspirations and practices in the field of technology on Biblical motives and quotations was common to Calvinist, Lutheran and Catholic authors alike. In this respect, Protestants such as Erhard Weigel or Georg Harsdörffer and Jesuits such as Athanasius Kircher and Kaspar Schott were all of one mind.[72]

Finally, the complexities of the relations between religious contexts and visions of nature in early modern Europe are vividly illustrated by the vicissitudes of a region in Germany. The Electorate Palatinate, straddling the River Rhine near Speyer and Mannheim, went through a long series of religious upheavals between the middle of the sixteenth century and the beginning of the eighteenth century. Following dynastic events and the fortunes of war, between c.1550 and 1700, the denomination of the established church of the region shifted from Catholic to Lutheran to Calvinist to Lutheran to Calvinist to (briefly) Catholic to Lutheran to Calvinist and back again to Catholic. The religious composition of the – mostly rural – population, meanwhile, was mixed: partly Protestant, partly Catholic. After about 1690, when the Electorate officially returned to the fold of the Church of Rome, the majority of its inhabitants remained Lutheran or Calvinist.[73]

Forests were an important natural resource for the Palatinate, both for its ruler and for the peasants. During the early modern period, the ruler steadily managed to strengthen his hold on the wilderness, to the detriment of the peasants. By issuing by-laws, mapping the territory and expanding the forestry service, the sovereign lord of the Palatinate succeeded in tightening his grip on the woodlands, in order to reserve their use mainly for hunting and for commercial purposes, such as the export of timber to the Low Countries. Access to the forests as a common resource for the locals became more and more restricted. The tendency towards the increased exploitation of natural resources under the aegis of the state accelerated markedly after 1700 – the very period when Catholicism once again enjoyed the status of official denomination of the realm. In this prolonged contest over the use of the wooded wilderness, the territorial ruler, his 'rational' bureaucracy and the clerical elites were aligned against the mass of the rural population, both Protestants and Catholics. Whereas the prince and his allies chiefly saw the forest in terms of exploitation,

---

72 Stöcklein, *Leitbilder der Technik*, 13, 25–26, 43–45.

73 Allmann, *Wald in der frühen Neuzeit*, 25–26, 362.

Protestant and Catholic peasants continued to envisage the woods as a separate, animated space, which should not be unduly infringed upon, on pain of sanctions by supernatural powers. Protestants and Catholic peasants alike resorted to magical practices to keep these fearsome forces at bay.[74] The dividing line in visions of nature and the use of resources within this thoroughly mixed territory was thus not primarily confessional, but social and political. Once again, differences in visions of the uses of nature between confessions in Latin Christianity hardly mattered.

Conclusion

This chapter examined several perspectives on differences in visions of the uses of nature between religions, and asked whether these differences, if any, mattered in practice for the way in which nature was approached. Time and again, the conclusion was negative. Unmitigated exploitation of nature was not the default position in Latin Christendom. In fact, the Bible allows for a variety of attitudes towards the natural world. There is no evidence that from the Middle Ages onwards, the majority of Christians in Europe were moved by Scriptural exhortations to seek a complete domination of nature rather than paying heed to alternative values, such as the importance of 'stewardship'. Moreover, images of domination were not a creation or a preserve of Latin Christendom. Ideas and practices conducive to the exploitation of nature could be found before the advent of Christianity and also existed in parts of the world where entirely different religious traditions reigned supreme – outside the orbit of Latin Christendom. Within early modern Europe, various Christian denominations, especially Calvinism and Catholicism, did not differ as much in values and practice regarding the treatment of nature as sometimes has been supposed. We may therefore, in short, conclude that variations in visions of the uses of nature between religious traditions between c.700 and 1800 have simply been too weak or too inconsistent to offer part of the explanation for the differences in technological development.

This conclusion has wider implications, which merit some further reflection. If differences in technological development were hardly connected to variations in visions on the uses of nature, the question arises whether it is even necessary to include such overarching mental frame-

[74] Allmann, *Wald in der frühen Neuzeit*, 46–51, 99, 185, 191, 262–263, 287–321.

works in an explanation of disparities in technical evolution and economic performance. A tacit assumption in debates about the relation between religion and technology appears to be that the explanation of technological change requires the pre-existence of general moral views or orientations that sanction or restrain the manipulation of nature. Technological change is supposed to have needed some sort of moral justification, religious or otherwise. However, even if it can be readily accepted that such overarching belief-systems exist – whether we call them 'cosmological beliefs', 'controlling imagery', or in a more-old-fashioned way, simply 'ideologies'[75] – and that these belief-systems powerfully affect the ways in which people interpret the world and view their own lives, this does not entail that the ultimate explanation for what people do should be sought at this deep level of causation. It is perhaps pushing reductionism too far to go all the way back to the substratum of worldviews in order to understand the evolution of technology.

The ways in which the impact of religious contexts on technological change have been approached up until now have not been, to my mind, singularly fruitful. Studies on the impact of religion on technological innovation starting from the side of religion have tended to get stuck in casuistic discussions on the interpretations of texts or images and have frequently been in want of a rigorous comparative perspective.

However, even if differences in visions of nature between religious traditions turn out to be irrelevant, this does not mean that variations between religious contexts *as such* are not important in explaining disparities in the evolution of technical knowledge. It only means that relevant explanatory variables should be sought at a level other than that of the sphere of ideas, attitudes and values; this latter angle leads to a dead end. Research strategies should be refocused, and answers may be found in a different direction from that which scholarly conventions might lead us to expect.

Differences in religious contexts, I would suggest, influenced the development of technology primarily by means of the ways in which they affected institutions and underlying socio-political structures for the formation of human capital and the circulation and creation of technical knowledge. Variations in religious contexts were most important at the level of institutions, patterns of communications and movements

[75] Lal, *Unintended consequences*, 7–8, Merchant, *Death of nature*, O'Brien, 'Needham question updated', 20–22.

of people connected with religious traditions. It is at this level that religion had an impact on technological change. The task of the following three chapters will be to show how those connections actually worked in the formation of skills, the circulation of knowledge and the creation of novel technologies. The first angle that we will take concerns the relation between religion and the formation of skills.

CHAPTER TWO

RELIGION AND HUMAN CAPITAL FORMATION IN CHINA AND EUROPE

Human capital is the stock of expertise acquired by an individual worker. Human capital formation is a shorthand term for an individual's increased capacity to produce, earn, spend and enjoy life. In Fritz Machlup's words:

> one speaks of formation of human capital when investments ... are made to transform the mental or physical equipment embodied in individual persons in ways that enable them (1) to produce more or better goods or services, (2) to earn higher pecuniary incomes, (3) to spend their incomes more intelligently, or (4) to get more pleasure out of life thanks to enhanced awareness and appreciation of some "finer things". Individuals can acquire and improve these capacities by investment in training or education.[1]

In unified endogenous growth theory, human capital is viewed as a key determinant in the transition to the 'modern growth-regime', characterized by reduced population growth and sustained income growth. The reasoning is that once technological progress, induced by an increased size of population under Malthusian conditions, has become sufficient, parents will find it optimal to provide their children with some human capital. Then a virtuous circle will emerge: higher levels of human capital increase technological progress (and economic growth), which in turn raises the value of human capital and the returns on investment in the training and education of children. Eventually, parents will spend more of their resources on 'child quality' and less on 'child quantity', which will in time result in a decline in fertility.[2]

How, and to what extent, did religious institutions contribute to the formation of human capital prior to the onset of the modern growth regime? The general argument of this book, as the Introduction explained, is that religion's chief point of impact on the development of technology should be sought at the level of institutions, patterns of communication and movements of people connected with religious traditions, rather than in the sphere of ideas, attitudes and values, where most students of the

[1] Machlup, *Economics of information*, 420–421.
[2] Galor and Weil, 'Population, technology and growth', 811, 819, 826.

subject still prefer to dwell. This chapter takes up the discussion specifically with regard to religious institutions in so far as these affected the formation of 'mental or physical equipment embodied in individual persons' that made them more capable in the use, improvement or innovation of technical artefacts and practices, in other words: human capital formation.

Human capital formation can be realized in various ways, namely via learning by doing, learning by using or formal learning. Contrary to the conventional assumptions made by some economists and economic historians, the use, improvement and innovation of technology in pre-industrial economies did not necessarily only take place through learning by doing or learning by using, made possible by a division of labour and specialization under conditions of a growth of aggregate demand. In pre-industrial economies, too, technological advance could be more than simply 'the outcome of optimally random implementing random discoveries', without the aid of any other mechanisms for transmission than mobility of skilled individuals and person-to-person transfer of knowledge.[3] Knowledge was not only embodied in humans, but also could also be incorporated in texts, images and artefacts. After all, the advent of literacy and numeracy and the spread of the printing press preceded the Industrial Revolution. Learning could and *did* take place in more formal, less personalized ways long before societies underwent an industrial transformation.[4] This chapter will give due attention to all of these possible modes of human capital formation.

Religious institutions will be understood here as durable organizational and spatial forms in which religious beliefs, values and practices find expression, such as churches, societies, orders, fraternities, charities and material places of worship, meditation or communal life. Material sites of religious institutions may include, for example, church buildings, chapels, temples, shrines, monasteries, convents and other types of religious houses. In order to find out whether, how, and to what degree religious institutions truly made a difference for the formation of human capital relevant for technology, I will follow a comparative approach. Comparisons will be made between various ways of learning, various religious institutions and different geographical areas. From this chapter onwards,

---

[3] Aiyar, Dalgaard and Moav, 'Technological progress and regress', 127, 129, Persson, *Pre-industrial economic growth*, 10–11.

[4] On modes of technological learning: Tunzelmann, *Technology and industrial progress*, 9, 119–121, 399–400, and Davids, *Rise and decline*, chapters 6 and 7.

as explained in the Introduction, the analysis will concentrate on comparisons between China and Europe as well as between different regions within Europe.

The comparative approach will consist of three steps. As a first step, I will give an overview of indicators on human capital formation and assess what they can tell us about variations in human capital formation in space and time. The next section of the chapter examines the role of religious institutions in China and Europe in the formation of human capital relevant for technology between about 700 and 1800 through 'formal' learning, i.e. learning through schooling in a teacher-pupil setting. In this section, I will argue against recent interpretations that ascribe human capital formation, such as the growth of literacy, primarily to the influence of Protestantism.[5] The final section of this chapter looks at the role of religious institutions in human capital formation through 'informal' ways of learning, such as learning by using and learning by doing, which resulted in tacit rather than explicit forms of knowledge. The conclusion addresses the key question, to what extent can variations in human capital formation relevant for technology in China and Europe be explained by variations in the spread and activities of religious institutions?

Variations in Human Capital Formation

Interest in human capital as an explanatory factor in long-term economic growth has increased in recent years. However, studying the evolution of human capital in history becomes more difficult the further one moves back in time before 1800. For large parts of the world, data to measure the stock of expertise acquired by individual men or women before the nineteenth century are not readily available, or even entirely absent. Evidence on training and education is less scarce, but provides, of course, only an indirect view of the capacities individuals may have obtained.[6] In the following, I will first review some of the indicators that historians use to estimate individual cognitive abilities in the period before 1800 and discuss their information on the Little and Great Divergence; next, I will turn to evidence on formal schooling; and finally, I will look at other

[5] For example, Becker and Woessmann, 'Was Weber wrong' and Huff, *Intellectual curiosity*, 309–310.

[6] A'Hearn, Baten and Crayen, 'Quantifying quantitative literacy', 783–784.

data on forms of learning by which individuals could extend their stock of knowledge and skills.

Until recently, the most common indicator used by historians to estimate the cognitive abilities of individuals in pre-industrial societies was that of literacy. Literacy is usually defined as the acquisition of some basic ability to read and write. As a proxy to measure the degree of literacy before 1800, historians typically looked at the ability of men or women to write their own signature in marriage records or in legal documents such as wills, deeds or depositions.[7] This measure is, of course, rather crude. Reading and writing abilities do not necessarily go together. Some people can read without having mastered the skill of writing and some people can write their own signature without being able to read. Moreover, signing one's own name is the simplest task of writing. It does not even automatically entail the ability to copy a phrase. But whatever its drawbacks as a measure for literacy more broadly, a signature is, at any rate, a direct form of evidence of an individual's writing ability.[8]

On the basis of signature data, Jaime Reis has presented a comparative survey of literacy in Europe at the end of the eighteenth century. This survey shows wide disparities within Europe, which did not correspond with religious divisions. The highest levels of literacy by 1800 – in the order of 60–80% among males and 40% among females – had been reached in Northwest Europe, i.e. Europe north of the Saint-Malo-Geneva and west of the Stralsund-Dresden line, which included both predominantly Protestant areas such as England, Scotland, the Netherlands and parts of North Germany, and staunchly Catholic territories like Belgium, Northern France and Western and Southern Germany. Regions showing lower levels of literacy – in the order of 10–45% among males and 10–20% among females – included both Lutheran Norway and Sweden and Catholic Italy, Portugal, Spain and Southern France.[9] Data are more sparse for earlier centuries. An informed guess by Harvey Graff puts the male literacy rate in Europe in the late fifteenth century at 5–10%.[10]

The extant evidence on signatures from the period between 1500 and 1800 gives an interesting glimpse as to how different regions in Europe

[7] Graff, *Legacies of literacy*, 4–7.

[8] The pros and cons are time and again discussed in the scholarly literature, see e.g. Houston, 'Literacy', 270–271, Graff, *Legacies of literacy*, 34–35, 435 note 161, Reis, 'Economic growth', 201.

[9] Reis, 'Economic growth', 201–203.

[10] Graff, *Legacies of literacy*, 106.

evolved during this time. Following strong growth throughout the sixteenth and seventeenth centuries, the Netherlands reached a male literacy rate of c.70% as early as 1700. Men's literacy in England doubled in the sixteenth century and again in the seventeenth century and grew by another 50% between 1700 and 1800. The disparity in literacy rates within France was already manifest at the end of the seventeenth century, with male literacy above the Saint-Malo-Geneva line reaching more than 30%, against less than 30% below it. During the eighteenth century both regions saw their levels steadily increase.[11] In Germany, literacy in Catholic regions in the seventeenth century was lower than in Protestant regions, but the Catholic regions started to catch up in the eighteenth century, reaching equality after 1800.[12] Literacy rates in Castile in the sixteenth and early seventeenth centuries rose as fast as those in England and France. Judging by signature rates in Inquisition trial records, male literacy rates in Madrid, Toledo, Cuenca, Santiago and Andalusian towns between 1600 and 1650 reached levels of 52–70%. It was not until the middle of the seventeenth century that stagnation set in.[13] Evidently, Catholic regions did not persistently lag behind Protestant ones in early modern Europe.

Signature data are not available for China. Estimates on the development of literacy in China before 1800 have been built on scraps of other kinds of evidence, such as information on the spread of elementary education and textbooks, the diffusion of printed materials and messages in writing or the size of various social groups, such as *literati*, clerks or merchants, who must have acquired some writing and reading skills. Evelyn Rawski estimated on the basis of this sort of evidence that in the mid- and late nineteenth century, 30–45% of the men and 2–10% of the women were able to read and write,[14] but other scholars have disputed this picture as being too optimistic. The expansion of elementary education under the Qing, for instance, had political, financial and cultural limits. The state, the *literati* and the value system in society were far from favourable to mass education.[15] Nevertheless, there is a consensus that *some* growth in

---

11 Reis, 'Economic growth', 202–203, Graff, *Legacies of literacy*, 192–217, 230–238, Stephens, 'Literacy', 550–560.

12 Hinrichs, 'Alphabetisierung', 545.

13 Nalle, 'Literacy', 68–69.

14 Rawski, *Education and popular literacy*, 23, 82, 140.

15 Woodside, 'Some Mid-Qing theorists', Leung, 'Elementary education', 381, Woodside and Elman, 'Afterword', 528–534.

literacy occurred during the Tang and the Song and again during the Late Ming and Qing, even if the extent of this increase is hard to determine.[16]

Another indicator, one that has been in vogue among economic historians in the past few years, is numeracy. Numeracy is the basic competency of quantitative reasoning, namely the ability to count, to keep records of one's counting and to make calculations.[17] As a proxy to measure the degree of numeracy before the nineteenth century, historians are now increasingly turning to data on 'age heaping'. Age heaping refers to the phenomenon of people self-reporting their age in an approximate but not random fashion, namely, their showing a systematic tendency to choose a figure ending in a five, zero or an even number. The frequency of this 'heaping' of ages is assumed to reflect a low ability of quantitative reasoning and thus a low degree of numeracy. Some historians claim that evidence on age heaping does not only provide an additional indicator of human capital, but that it also, because of observed correlations between age heaping and literacy, has the potential to extend knowledge of human capital *as such* to times and places where data on literacy are entirely absent or extremely scarce.[18] 'Age heaping' would thus vastly expand the possibilities for comparative research on human capital.

Caution is needed, however. Evidence on age heaping as an indicator of numeracy should be handled with at least as much care as evidence on signatures. The proposition that a high degree of age heaping in pre-industrial societies reflects a low level of numeracy, which underlies the far-reaching claims about age heaping as an indicator of human capital in general and numeracy in particular, still has the status of a heroic assumption rather than of a soundly established fact. No one has as yet demonstrated for any society in the past that an individual with a lack of age awareness has a low degree of ability in quantitative reasoning – and the dearth of historical records may well make that assertion impossible to prove.[19] In fact, it is simply not known whether a person whose age appears to be reported accurately is more adept in making calculations

16 Zürcher, 'Buddhism and education', 54–55, Pei-Yi Wu, 'Education of children', 308–309, 318, Woodside and Elman, 'Afterword', 530–531.

17 Emigh, 'Numeracy', 653, A'Hearn, Baten and Crayen, 'Quantifying', 785, Van Zanden and De Moor, 'Van fouten kan je leren', 56–57.

18 A'Hearn, Baten and Crayen, 'Quantifying', 805–806.

19 Cf. A'Hearn, Baten and Crayen, 'Quantifying', 788; moreover, there is seldom any confirmation from independent sources as to whether a reported age was accurate or not. Evidence of a high degree of accurary of reported ages in Leiden (the Netherlands) in the 18th century is provided by Davids, 'Migratiebeweging', 140.

than someone who gives his or her age in a figure ending in a five, a zero or an even number. 'Inaccuracies in age reporting may not necessarily reflect lack of numeracy per se; it may also reflect a somewhat narrower lack of biological age. Knowledge of assets may have been more developed, for example, than knowledge of biological age', Rebecca Emigh reminds us in her detailed study on numeracy in fifteenth-century Tuscany.[20]

High accuracies in age reporting, on the other hand, are not necessarily an indicator of a high ability in quantitative reasoning. They may also reflect different techniques in memorizing birth dates. Jowett and Li found that the high quality of age reporting among Han Chinese in the Chinese census of 1982, in contrast with that among other ethnic groups, could be explained by the widespread use of the traditional Chinese calendar, which made it relatively easy to recall the exact date of birth. A similar degree of accuracy is shown by other Asian populations who use the Chinese calendar.[21] Moreover, a strong correlation between levels of literacy and accuracy of age reporting does not hold at all times and in all places. No such relation could be found in data on Han Chinese and other Asian populations between the 1950s and 1980s.[22] Moreover, low levels of age heaping were also interrelated to marital status. Married women were less prone than widows or spinsters to round off their age to a multiple of five. The most plausible explanation for this phenomenon appears to be that 'wives adjusted their reported ages to that of their spouses'.[23]

The most extensive dataset on numeracy based on age heaping is that which has been assembled by A'Hearn, Baten and Crayen. This dataset refers to 130 places in all parts of Europe and covers the period between about 1350 and the beginning of the nineteenth century. Relying on this great variety and amount of data, A'Hearn, Baten and Crayen feel confident in drawing conclusions both about the evolution of numeracy and human capital in Europe in the very long run, and about variations in numeracy across Europe over time. They infer that 'basic mathematical skills' diffused earlier than literacy and that this process, prompted by the 'spread of market activity', started earlier in Western Europe and the German-speaking parts of Central Europe than in the regions lying further to the East. While the 'West' saw a 'slow but steady improvement' from the Late Middle Ages, resulting in numeracy levels of c.70% by 1600, the 'East'

---

20 Emigh, 'Numeracy', 670.
21 Jowett and Li, 'Age-heaping', 429–431.
22 Jowett and Li, 'Age-heaping', 433.
23 Földvári, Van Leeuwen and Van Leeuwen-Li, 'How did women count', 305, 310.

did not go through a similar process until after 1650.[24] Considering the methodological observations made above, however, it is not evident that the data presented reflect more than differences in accuracy of age reporting. Whether numeracy spread before literacy, and whether this occurred earlier in western regions of the Continent than in the eastern parts, is not established beyond dispute.

It is perhaps more revealing to note what the data do *not* reveal. Differences in accuracy of age reporting – and in the levels of numeracy that supposedly went with it – did *not* coincide with differences in religion. Weberian stereotypes do not hold. Rates of accuracy in age reporting had already gone up in North Italy (from 31 to 55%) and the Netherlands (up to 72%) before 1450 – long before Calvinism appeared on the stage. In 1600, Catholic Austria boasted a higher level (84%) than Protestant England (76%) or largely Protestant Bohemia (44%). Bohemia improved its rate to 85% in 1700, after its return to Catholicism, and France saw its rate increase from 79% in 1650 (during the heyday of religious pluralism under the Edict of Nantes) to 89% in 1700 (after the Huguenots had been expelled). Lest anyone were to think that Weber might have been right had he made the opposite claim, it may be pointed out that rates of accuracy in age reporting in Protestant parts of Germany in 1700 were higher than in Catholic regions: 87% against 68%.[25] It is apparently difficult to square age heaping habits with the Little Divergence.

What about the Great Divergence? Alfred Crosby has claimed that the rise of quantitative reasoning began earlier in Europe than anywhere else in the world. According to his thesis, the first stirrings of this transformation, which would eventually enable Europeans to make rapid advances in science and technology, could be detected between about 1250 and 1350. The essential change took place in what Crosby called the *mentalité*: collective habits of thinking about reality. From the late thirteenth century onwards, Europeans began to approach reality in ever more quantitative terms.[26] Crosby believed that this shift was apparent in a variety of areas, ranging from perceptions of time and space to the rise of perspective painting, the introduction of new techniques for notating music, the emergence of double-entry bookkeeping and the spread of marine charts provided with grids of rhumbs.

---

[24] A'Hearn, Baten and Crayen, 'Quantifying', 805–806.
[25] A'Hearn, Baten and Crayen, 'Quantifying', 801, table 4.
[26] Crosby, *Measure of reality*, xi, 18–19, 50.

Although Crosby declared that people in Europe 'were thinking of reality in quantitative terms with greater consistency than any other members of their species',[27] he did not probe more deeply into the diffusion of this quantitative approach within European society, nor did he make extended comparisons between Europe and any other region of the globe. In fact, the emergence of quantification in Europe may have started much later than Crosby claimed. Patricia Cline Cohen suggested that the rise of quantification in Europe did not take place until the seventeenth century, and she saw its appearance not as a harbinger of advances to come but as a response to economic, social and cultural changes that were already taking place, such as the expansion of ocean shipping, the growth of vagrancy and the increasing sensitivity to the wastage of mortality.[28] The evolution in mentality thus succeeded a transformation in reality.

Moreover, an inquiry in China, as one of Crosby's critics pointed out,[29] would easily yield instances of a similar kind of quantitative approach to that which Crosby found in Europe. The implicit message of Crosby's thesis, namely that on this dimension of people's cognitive abilities, Europe and China began to diverge much earlier than with economic development in general, has as yet not been corroborated by evidence. Data that would allow comparisons of numeracy in China and Europe are still lacking. Even material on age heaping in China before 1800 has not yet been produced. One may nevertheless hazard a guess as to what figures on accuracy of age reporting in China in this period would show. If Jowett and Li's conclusion about the impact of the use of the traditional Chinese calendar is right, it may be expected that the rate of accuracy among Han Chinese has been high for centuries. If A'Hearn, Baten and Crayen's suggestion about the influence of the spread of market activity is correct, too, the level of accuracy must have gone up by leaps and bounds from at least the Song onwards. China, after all, saw a remarkable rise in the number of market towns during the Song dynasty and again under the late Ming and Qing.[30] The extent of the development of commerce and credit mechanisms in Song times is said to have been hardly conceivable without 'a minimal necessary level of general literacy and knowledge of arithmetic'.[31] It is, alas, not yet possible determine whether these conjectures are true.

---

27 Crosby, *Measure of reality*, xi.
28 Cohen, *A calculating people*, 44–45.
29 Hart, 'The Great Explanandum'.
30 Von Glahn, 'Towns and temples', 177, 181–187.
31 Pei-Yi Wu, 'Education of children', 308–309.

Evidence that throws a more direct light on the development of numeracy in China, as distinct from schooling, has not yet been uncovered.

Considering the shortcomings of traditional measures of human capital, such as the ability to write one's signature, Joerg Baten and Jan Luiten van Zanden have proposed using book production per capita as a better alternative. They claim that the number of new and re-edited printed books per capita in a given society reflects aspects of human capital formation that are not adequately covered by measures such as signature rates. 'Book production' and 'book consumption' can, in their view, be used interchangeably. Analysing a broad set of data on estimated total book production in eight European countries (Great Britain, Germany, France, Italy, Spain, the Netherlands, Belgium and Sweden) in relation to a proxy of per capita income between c.1450 and 1800, Baten and Van Zanden concluded that 'it was exactly in the countries in which book production increased fastest that real wages developed systematically better over the centuries before the Industrial Revolution than in countries with lagging human capital formation'. The forerunners were the Netherlands and Britain and the laggard was Spain, while the other countries formed a broad echelon in the middle.[32]

In a follow-up study, Van Zanden and Buringh broadened the comparison of book production both in time and in space. They extended the dataset on book production in Europe to include estimates of all individual manuscripts produced between 500 and 1500 and of all new titles or editions of books printed between 1454 and 1800 in the whole of Western and Central Europe, and supplemented these with estimates on average print runs. Next, they briefly compared book production in Europe before 1800 with some recent figures on output of books in China in Ming and Qing times and in Japan during the eighteenth century. They concluded that book production in Europe increased, with ups and downs, from the sixth century onwards. Growth was particularly rapid during the Carolingian Renaissance of the eighth and ninth centuries, the High Middle Ages from the eleventh to the thirteenth century, and during the period after c.1450, when the emphasis in production changed from script to print. During this long period of expansion, the centre of gravity repeatedly moved from one region to another. It shifted from Italy to northern France, Belgium and the Rhineland (with Ireland, Britain and Spain), then southwards back to northern Italy and Switzerland and finally northwards

32 Baten and Van Zanden, 'Book production', 218, 220–221, 232.

again, to the Netherlands, England and France. Levels of book production in Europe after 1500 appeared to be higher than in China and Japan, which – according to Van Zanden and Buringh – might imply that levels of human capital were also higher in Europe, which might help to explain the Great Divergence.[33]

Book production per capita has the significant advantages of quantifiability and comparability. Yet, the use of this material as an indicator of human capital is also beset with difficulties that have as yet not been fully resolved. In contrast with signatures, book production per capita is an aggregate, indirect measure rather than an individualized, direct measure of a person's ability. While the advocates of this new indicator confidently claim that Spain, despite its 'surprisingly high' level of literacy up to 1650, apparently did not possess the 'more advanced components of human capital' because printed book production per capita was relatively low,[34] they do not explain what the basic components of human capital measured by their indicator actually are.[35]

What exactly does output of printed books per capita tell us about individual cognitive abilities? To know the answer to this question, it would be necessary to have information about what sort of knowledge or skills the new or re-edited books could transmit and which people were reading or consulting these books. The output of printed books in any society at any given time comprises very diverse materials, which do not always have an obvious connection with the formation of human capital in the first or second sense distinguished by Fritz Machlup, but that perhaps *do* reflect it in the last sense he mentioned: to get more pleasure out of life, thanks to enhanced awareness and appreciation of 'finer things'. The vast majority of books in the *bibliothèque bleue*, for instance, which found their way all over the countryside of eighteenth-century France, consisted of songs and carols, bible stories and scripture, lives of saints, religious tracts, fairy stories, tales of chivalry, novels and stories and plays and morals.[36]

---

33 Buringh and Van Zanden, 'Charting the "Rise and the West"', 412–414, 436–438, 440–441.

34 Baten and Van Zanden, 'Book production', 221; in Buringh and Van Zanden, 'Charting the "Rise and the West"', 433–434, the discrepancy is resolved by calculating rates of literacy based on estimates of per capita book consumption, which results in very low figures for Spain before 1800.

35 Cf. Nuñez' remarks on the importance of desaggregating the concept of human capital, Nuñez, 'Literacy', 540–541, 542–543.

36 Graff, *Legacies of literacy*, 218–219.

Moreover, printed book output per capita is not a very refined indicator of individual book consumption in a given society. Although Baten and Van Zanden use the terms 'book production' and 'book consumption' interchangeably, presumably on the assumption that the phenomenon of re-editions is sufficiently revealing about demand, in reality, book production per capita does not tell us much about the consumption side. Some of the books produced in a country may have been destined for export markets, whereas some of the books that were actually 'consumed' may have been imports.[37] Part of the production may never have been sold at all. Neither do figures on book production give us insight into the spread and use of books in a particular society. They give no information about ownership and readership. Who were the owners of books? Who had other forms of access to books? What did people do with those books?[38] Or to put it more succinctly: supposing that book production is an indicator of human capital, how was this capital distributed in a given society? It seems to me that from the point of view of human capital formation, such questions are even more urgently in need of an answer than queries about the total number of books produced.

Finally, one should also keep in mind that even after the spread of printing with movable type, printed books were not the only books in circulation after 1500. Books in manuscript form continued to be produced and distributed. Notably, in Spain in the early modern period, many texts circulated as manuscripts rather than in print, one of the reasons being that manuscripts more easily escaped censorship by the state or ecclesiastical authorities. The number of copies of a printed book was not necessarily higher than that of a book reproduced in script.[39] Against this background, it seems somewhat premature to infer from a relatively low printed book production per capita that Spain did not possess 'more advanced components of human capital'.

Books may reveal something about a divergence between China and Europe, but not in any easily measurable way. The main line of the evolution of book production in China was not unlike that reconstructed for Europe by Van Zanden and Buringh. A first, modest wave of growth

[37] The appendices in www.iisg.nl/bibliometrics do not clarify sufficiently how the issue of exports and imports has been resolved, except by estimates based on book fairs in Germany.

[38] Cf. also the pertinent remarks by Stephens, 'Literacy', 548–549.

[39] Bouza, *Corre manuscrito*, 16–19, 31, 35–38, 63, 66–67, 82–83, Portuondo, *Secret science*, 63–64, Love, *Scribal publication*, 1–2, 32–34, 177–178, 184–185, Chartier, *Inscrire et effacer*, 117–118.

between the eighth and tenth centuries, under the Tang dynasty, was followed by a second, greater spurt between the late eleventh and late thirteenth centuries under the Northern and Southern Song, a third burst of expansion of massive proportions during the sixteenth and early seventeenth centuries under the Late Ming, and another wave of expansion under the Qing from the end of the seventeenth century onwards.[40]

However, the size of the total output remains hard to establish. The number of printed books issued in China between c.1450 and 1600 was, according to Joseph McDermott, 'certainly far, far lower' than that estimated for Europe in the same period, and the annual number of new titles published in during the Late Ming definitely lagged behind that produced in Europe.[41] Average print runs in China were probably smaller than in Europe, too.[42] But such estimates do not say much about the actual size of the production of books. Even though in China from the sixteenth century onwards, starting in the Yangzi Delta, the printed book gained ascendancy over the manuscript, the manuscript tradition remained alive and well until the late twentieth century. Many books were repeatedly copied in manuscript form and some genres, such as genealogies or instruction manuals on family practices and social customs, continued to circulate in manuscript form only.[43] In the Chinese printing industry, movable-type printing grew from the fifteenth century onwards (both with wooden and metal types), but woodblock printing continued to be the prevalent technology. Woodblock printing was more flexible than printing by movable type, which after 1450 quickly became the dominant technology of book production in Europe. As the capital requirements for woodblock printing were low compared to printing by metal movable type, and the need to produce large runs was reduced accordingly, the woodblock method allowed printers to print books in instalments or to produce copies 'on demand'. Unsold copies were much more rare in China than in Europe. The scale and speed of production could simply be adjusted to demand. Additional carvers could easily be hired when and where the need arose and woodblocks, once carved, could be used time and time again.[44] It is

---

40 Chia, 'Machaben', 286–307, Chow, *Publishing*, 20–23, McDermott, *Social history* 43–78, Brokaw, *Commerce in culture*, 8–19.

41 McDermott, *Social history*, 70–71, Bray, *Technology and society*, 11–13, Chia, *Printing for profit*, 7–13, Chia, 'Mashaben', 303–304, Buringh and Van Zanden, 'Charting the "Rise and the West"', 437.

42 Chow, *Publishing*, 248.

43 McDermott, *Social history*, 76–78, Blair, 'Afterword', 350.

44 Chow, *Publishing*, 59–71, 247–249, McDermott, *Social history*, 44–54, 79–81.

said that depending on the quality of the wood and the carving, a single woodblock could be used for 15,000, 20,000, 30,000 or even 40,000 impressions.[45] Thus, the size of an average print run does not give a very accurate idea of the total number of copies produced.

As the number of unsold copies of printed books in China was lower than in Europe and book production was primarily oriented to the domestic market,[46] production estimates for China before 1800 may reflect the size of inland consumption better than figures for separate countries in Europe. The upsurge in output from the late fifteenth century clearly met a growing demand. Unlike the production spurt during the Song, most of the printed material under the late Ming and the Qing was no longer government publications but books issued by commercial publishers. The share of government imprints in total production declined from nearly 50% between the tenth century and the late fifteenth century to merely a quarter in about 1570.[47]

Moreover, the nature of the output became much more varied. Next to books in traditional categories such as the classics, Confucian works or medical texts, a large part of the output now consisted of other sorts of publications, such as household encyclopaedias, divination texts, anecdotes, stories and similar entertainment lore, which would have not have been unfitting in the *bibliothèque bleue*.[48] Commercial publishers, according to Timothy Brook, 'went well beyond' the 'Confucian canon and the standard histories as well as the *The Ming Code* and *Ming Regulation*... producing all manner of popular texts cheaply and in large volume to sell to a broader reading public than aspiring officials'.[49] As in Europe, a sharp drop in prices and a vast expansion of distribution networks brought books within the reach of a larger group of customers. Kai-Wing Chow showed that the relatively low prices of books in the Late Ming meant that books 'were affordable not only to the gentry, merchants and officials, but also to ordinary urban workers'.[50] An extended network of bookshops, bookstalls, fairs, peddlers and temple grounds, which reached a particularly high degree of development in the Lower Yangzi area, provided the channels along which books could find their

[45] Chow, *Publishing*, 34, McDermott, *Social history*, 21.
[46] Chow, *Publishing*, 248–249.
[47] McDermott, *Social history*, 68–69.
[48] Chia, 'Mashaben', 307–312, McDermott, *Social history*, 62–69, Brokaw, *Commerce in culture*, 523–533, 554.
[49] Brook, *Confusions of pleasure*, 131.
[50] Chow, *Publishing*, 47–56.

way from publishers to buyers. This trend spread to other areas in the empire under the Qing.[51]

The indicators discussed so far primarily relate to the acquisition of cognitive abilities, and in particular to the understanding and manipulation of symbols such as letters, characters or numbers. As a measure of human capital formation, these indicators have their uses but also their limitations. One of the limitations is the inherent, but unproven, assumption that the abilities to read, write or calculate must have been as important in the formation of 'human capital' for people in the past as they appear to us today. Did human capital formation always require the acquisition of literacy or numeracy, or could it perhaps also be realized through the acquisition or cultivation of other cognitive abilities? Fernando Bouza has argued that in sixteenth and seventeenth-century Spain, hearing, seeing and reading/writing were seen as equally valid ways of knowing and making known. A fixed hierarchy between voice, image and written word did not exist; which of these ways of communicating and memorizing was preferred depended on the circumstances.[52] In addition, David Cressy has reminded us that in early modern England, 'a competent and contented life could be lived entirely innocent of literate skills'. 'Literacy was not necessarily as valuable or as enlightening as is often claimed', and low literacy rates in the past should therefore 'not be taken as indication of retardation or deprivation'.[53] In view of the different aspects of human capital formation distinguished in Fritz Machlup's definition, this makes perfect sense.

Another limitation of the indicators discussed so far is their exclusive focus on abilities that somehow found expression in grasping and using formal terms or symbols. An indicator that can be used to measure human capital more broadly than literacy, numeracy or book production/consumption is the so-called 'skill premium': the relative wage of skilled to unskilled workers. It reflects the extra remuneration for the acquisition of skills. Skills can comprise a various and varying set of abilities, ranging from cognitive abilities to tacit know-how to extract, shape or handle materials and material objects. The skill premium is determined by a variety of factors, including supply factors such as the costs of training, interest rates and the period during which a skilled worker can earn a

---

51 McDermott, *Social history*, 94–103, Brokaw, *Commerce in culture*, 19, 548–553.
52 Bouza, *Comunicación*, 31, 38.
53 Cressy, 'Literacy in context', 306, 309.

higher wage, as well as demand factors such as technological change and demographic developments.[54] Thus, the skill premium can partly, but not exclusively, be regarded as an indicator of remuneration for investment in human capital. A reduction in the skill premium is often assumed to reflect an increase in investment in human capital.

Nowadays, economists and economic historians are engaged in estimating skill premiums in various parts of the world in the very long run. On the basis of a dataset of wages of skilled and unskilled workers in the construction industry, Jan Luiten van Zanden has recently ventured to chart the long-term evolution in skill premiums in a broad range of cities in Europe and Asia between the fourteenth and nineteenth centuries and to explain the patterns of divergence and convergence that appeared to emerge. In Europe, skill premiums generally showed a sharp decline after the Black Death, but began to diverge from the middle of the fifteenth century onwards. While the skill premium in cities in England, the Low Countries and northwest France more or less remained at the relatively low level reached in about 1450, the skill premium in cities in northern Italy, Castile and Poland tended to rise again. Cities in the southern part of the Holy Roman Empire (Strasbourg, Augsburg, Vienna) after 1450 displayed approximately the same pattern as those in the North Sea area.[55] Asia showed wide disparities, too. Skill premiums in India, Java and Korea in the eighteenth and nineteenth centuries were much higher than in Japan and China. Within China, skill premiums in the region around Beijing in about 1770 were almost twice as high as those in the South. Levels in North China were comparable to those in other parts of Asia, whereas those in the South were on a par with those in Northwest Europe.[56]

Although Van Zanden concluded that low skill premiums generally reflected a high level of human capital formation and that 'institutions for the formation of human capital [were] working well', it is not entirely clear that such inferences are fully warranted by the extant data.[57] After all, the relative weight of various supply and demand factors that affected the level of the skill premiums has not yet been firmly established. We cannot be certain about the extent to which a low skill premium in a particular case was the result of investment in human capital rather than of, say, a higher level of remuneration for unskilled labour. In North Italy, for

[54] Van Zanden, 'Skill premium', 123–124.
[55] Van Zanden, 'Skill premium', 127, 129.
[56] Van Zanden, 'Skill premium', 131–133.
[57] Van Zanden, 'Skill premium', 147.

instance, the renewed rise of the skill premium after 1450 was apparently not merely linked to supply-side factors but also to demand-related factors, such as population growth. In Japan, the wage rates of unskilled workers in cities were pushed up by the relatively high level of labour productivity in agriculture, because the urban and rural labour markets for unskilled labour were closely linked.[58] Rising labour productivity in agriculture in South China – as contrasted with the lingering growth in the North – may have had a similar effect on the wage levels of unskilled labour (and the skill premium) in this region of China in early Qing times.[59]

The picture that emerges from this survey of indicators on human capital formation is diverse and complex. The nature, timing and spatial patterns of the growth, stagnation or disparities in the formation of human capital are far from simple and straightforward. In Europe by 1800, indicators such as literacy, accuracy of age reporting, book production and skill premiums show outstanding scores for England, the Low Countries and northern France. But such indicators, as we have seen, cannot necessarily be considered as accurate reflections of human capital formation. Moreover, the positions of different regions on these indicators turn out to have shifted and diverged in the course of time. During the Late Middle Ages, northern Italy led the way in book production and spearheaded the growth in accuracy of age reporting. Castile boasted relatively advanced rates of literacy until well into the seventeenth century. Skill premiums showed similar patterns across Europe until about 1450, when disparities began to emerge between Northwest and Central Europe on the one hand, and parts of Southern and Eastern Europe on the other.

The extant data on these same indicators further suggest that trends in human capital formation in China were not dissimilar to those in Europe. As in Europe, literacy in China appears to have been growing slowly prior to 1800. Skill premiums in South China by the late eighteenth century were comparable to those in Northwest Europe, while premiums in the North were more similar to those in other parts of Asia. As in Europe, book production grew by leaps and bounds from the eighth century onwards, showing an especially vigorous expansion from the end of the fifteenth century. The *timing* and *direction* of the trends thus do not seem to have greatly differed. Differences between Europe and China may still

58 Van Zanden, 'Skill premium', 141, 143.

59 Cf. Li, 'Was there a "fourteenth-century turning point"', 170, and *idem, Agricultural development*, 12, 138–141 on rising labour productivity in Jiangnan from the Song until the mid-nineteenth century.

have resided in the *level* of the trends as well as in the pattern of distribution of human capital in society. At the present state of inquiry, however, the extent of these differences cannot be captured in quantitative terms.

Religious Institutions and Formal Learning

Having conducted an overview of the state of research on human capital formation, let us now turn to the role of religious institutions. The key question is: did religious institutions make a difference in the formation of human capital relevant for technology? Was the formation of technology-relevant human capital in Europe and China between c.700 and 1800 influenced by religious institutions, and if so, how and to what degree?

Regarding the rise of literacy and numeracy, Rebecca Emigh mentions three conventional explanations, which she subsumes under the heading 'top-down-explanations': formal education, market capitalism and government bureaucracies. The common denominator of these explanations is the assumption that forces from 'above' somehow encourage or compel people to acquire the abilities to read, write or count, or at least create the circumstances that make the acquisition of such skills eminently desirable. Top-down explanations, as Emigh and others argue, are not wrong but incomplete. They give only a partial view of the process that led to a growth in individual cognitive abilities.[60] 'Schools were the principal but by no means the exclusive vehicle for the spread of literacy', Jaime Reis reminds us.[61] Clara Nuñez observes that 'schooling, or acquiring literacy, seems to be only a first step in the building of a human capital stock, on-the-job training and out-of-school programs being relevant to the literacy levels of adults'.[62] Emigh makes a similar point about numeracy: 'numeracy and the use of quantitative categories are part of everyday life. Ordinary individuals display considerable skill at generating problem-solving skills independently of formal education or formal representational systems'.[63] What holds for literacy and numeracy, I would add, is all the more relevant for those abilities that rely to a greater degree on 'tacit' knowledge, such as skills in extracting, shaping or handling materials and material objects. For such abilities, informal learning is at least as

[60] Emigh, 'Numeracy', 656–658.
[61] Reis, 'Economic growth', 203.
[62] Nuñez, 'Literacy', 538.
[63] Emigh, 'Numeracy', 660.

important as learning via formal education. The argument can equally be extended to other forces from 'above' that are said to propel the rise in literacy and numeracy: market capitalism and government bureaucracy. The 'top-down' explanation neglects the possibility of agency by individuals from other groups than economic or political elites.[64]

Rebecca Emigh therefore proposes to complement the top-down approach with a bottom-up view, which emphasizes the everyday practices of individuals and the way these practices, in interaction with states' policies, served as prime movers in the growth of numeracy (and other individual cognitive abilities). More specifically, her detailed case-study of rural Tuscany in the Late Middle Ages shows that states can take advantage of pre-existing levels of numeracy in a given society (for example, for collecting data for tax purposes) rather than creating numeracy in the first place, and that numeracy can spread and develop when local markets for land, labour and commodities arise, even if these markets are not (yet) fully capitalist. Thus, neither state bureaucracies nor formal education nor market capitalism can always be regarded as the prime driving forces of a rise in numeracy.[65]

The common top-down explanations and the new bottom-up approach suggested by Emigh do not explicitly take account of the role of religious factors. Patricia Cohen's pioneering study on numeracy in early America discussed the possible interrelations between numeracy and religion but discovered no evident links between the two. However, her discussion did not probe very deeply or widely. Cohen restricted herself to analysing the worn-out theses on Calvinism and commerce, or Puritanism and science, as potential connecting mechanisms between religion and numeracy, and found them wanting.[66] Connections at other levels than in the thin atmosphere of attitudes and values were not explored, and religious institutions were not considered.

Human capital formation can thus occur in both formal and informal settings. Formal learning is certainly not the only channel though which an individual can acquire particular abilities that are relevant for technology. Any explanation of human capital formation must therefore take account of a variety of factors, which may partly act top-down and partly operate bottom-up. Learning in informal settings, and the function of

64 Emigh, 'Numeracy', 657–658, 689–690.
65 Emigh, 'Numeracy', 655–658, 689–690.
66 Cohen, *A calculating people*, 41–43.

religious institutions therein, will be the subject of the next section of this chapter. In this present section, I will examine the role of religious institutions in formal learning and relate this to the different 'top-down' and 'bottom-up' forces discussed above.

Schooling for general education started long before schools were created for purposes of technical or vocational training. Formal learning via schooling for general education came into being once knowledge and skills began to be transmitted via teacher-pupil relations outside the workplace. The origins of this process in China and Europe reach back to a time well before the period covered by this book. In this section, I will take a comparative view of the evolution of formal learning, from the seventh century AD until the beginning of the nineteenth century. I will argue that the role of religious institutions in the evolution of formal learning in China and Europe swung repeatedly between divergence and convergence until about 1500, but finally started to diverge thereafter. Durable differences began to appear between China and Europe and within Europe itself. These differences, which intensified from the seventeenth century onwards, found expression in the extent and nature of education in general, as well as in the timing and degree of development of educational provisions aimed at providing technical and vocational training. The roots of these disparities lay in the varying relations between governments, markets, religious institutions and demand 'from below'. As a consequence, by the end of the eighteenth century, as contrasted with the situation in about 1100, religious institutions in some regions of Europe weighed much more heavily in the provision of general education and of vocational and technical training than in other parts of the Continent. More generally, they enjoyed a greater impact on formal learning in Old Regime Europe than in contemporary China.

From Divergence to Convergence and Back Again

Under the early Tang, China possessed a more extensive system of formal education than the world of Latin Christendom. In the first half of the seventh century, the Tang government introduced an elaborate network of official schools to generate a steady flow of candidates for offices in the imperial bureaucracy. The primary purpose of these schools was to educate students in the Confucian tradition in preparation for examinations that would admit them to the ranks of scholar-officials. The apex of this system was the state academy directorate in the capital, the 'Directorate of National Youth', which comprised the Imperial University, great schools for instruction in Confucian canonical texts and specialist schools for law, calligraphy and mathematics. In the provinces, schools were created at the levels of prefectures and counties. Every one of these local government schools had to build a Confucian temple. State-founded schools only provided middle and higher education, however; education at an elementary level was not offered. Moreover, the official school system largely collapsed during the devastating An Lu-shan Rebellion in the 750s and did not completely recover thereafter. Although the state academy directorate more or less resumed its functions, the network of schools in the provinces was only patchily restored. While some prefectures or counties saw their schools rebuilt, many others were forced to do without them.[67]

Outside the official school system, meanwhile, various facilities for formal education emerged, which to some extent complemented the provisions offered by the state. The introduction of civil service examinations as an admission ticket to a career as scholar-official evidently put a premium on the acquisition of a particular set of knowledge and skills and made it patently worthwhile for parents to try to provide their children with a measure of apposite human capital. Alternative educational facilities arose partly 'from below', partly from the growing number of Buddhist institutions spreading across China. From the Mid-Tang onwards, clans often set up their own schools to prepare family members for state examinations. Schools for this purpose were also started by private teachers, who admitted whoever was prepared to pay to study, or by wealthy families, who 'might open [them] to poorer neighbours'.[68] Furthermore, there is ample evidence from the latter half of the Tang to suggest that Buddhist monasteries not merely trained their own members, but also played a role in the education of laymen. Buddhism in China mushroomed after c.500 and reached its zenith in the later eighth and ninth centuries. The number of monasteries, temples and shrines soared and the ranks of the monks and nuns vastly swelled, with increasing numbers of clergy recruited from families of high official status. When Emperor Wu Zong commenced a frontal attack on Buddhist institutions in 845 (the attack was abandoned a few years later), the Buddhist establishment comprised some 4,600 monasteries and 40,000 smaller religious sites, and counted at least 260,500 fully ordained monks and nuns and probably between 600,000 and a

[67] McMullen, *State and scholars*, 17–18, 35–42, 52–58, Lee, *Education*, 71–73, 75–77, 81–82, Zürcher, 'Buddhism and education', 39, 43.

[68] Lee, *Education*, 76, Lee, 'Sung education', 106, McMullen, *State and scholars*, 48.

million religious in all. The total size of the Buddhist clerical body at this time is estimated at 1.5–2% of the entire population of the empire.[69]

Although not every Buddhist institution was engaged in external educational activities to the same extent, the extant evidence suggests in any case that, especially in the ninth and tenth centuries, monasteries 'throughout almost the whole of China' offered elementary education to boys from age groups that were not eligible for the official schools and/or provided instruction to young people who aspired to sit the entrance examination for the imperial bureaucracy. Pupils and teachers presumably either met in a village school connected to a monastery or in a monastery compound itself. The primers and other materials used in these educational activities for the laity appear to have been at least partly of a non-Buddhist nature. Buddhist monasteries also became vibrant centres of reproduction of religious texts, both for internal use and for distribution among customers from outside. The Chin-kuang-ming monastery, for example, employed a staff of 55 copyists for this purpose, including 28 monks and 27 laymen.[70]

Texts were not only copied in script; Buddhist monasteries were also early centres of printing. The oldest extant printed book, a copy of the Diamond Sutra from 868, was found among the documents at the Dunhuang monastic complex in Northwest China.[71] In this way, religious institutions in Tang China made a further contribution to the development of formal education.

After the Northern Song had restored and extended the civil service examination system and had reopened the Imperial University in the capital, in the course of time, the government also took steps to rebuild the lower tiers of the official school system. During the great waves of educational expansion and reform in the 1040s and 1070s, the Song authorities founded scores of schools at the prefectural and county levels. These local official schools, which were not seldom founded on sites where Confucian temples of former Tang schools were still standing, admitted children from both poor and wealthy families. Some of these schools also provided elementary education.[72] Yet, the government-sponsored supply

---

[69] Zürcher, 'Buddhism and education', 23–28, Gernet, *Buddhism in Chinese society*, 6–8, 12, 318–319 note 27, Lee, *Education*, 76–77.

[70] Zürcher, 'Buddhism and education', 43–50, Lee, 'Sung education', 105–106.

[71] Zürcher, 'Buddhism and education', 51–55, Kieschnick, *Impact of Buddhism*, 181, McDermott, *Social history*, 11.

[72] Lee, 'Sung education', 108–120, Lee, *Education*, 81–82, Kuhn, *Age of Confucian rule*, 126–128.

of educational provisions never entirely satisfied the increase in demand. The state school system did not begin to expand until a few decades after the re-establishment of the examination system and the network did not become very dense. Plans for founding government schools were not always fully realized in practice.[73] A manifest indication of the growing thirst for education among the population was an increase in autodidactic practices in the early Song. A sharp drop in the price of printed books between the ninth and eleventh centuries allowed many more people to prepare for the examinations by studying on their own.[74]

Just as under the Tang, the increased demand for education was largely met by a diverse assortment of facilities outside the official school system. Buddhist establishments, which by 1220 included some 400,000 monks and 61,000 nuns, or about 1% of the total population of the Southern Song Empire, continued to offer educational opportunities both for laymen and for clergy.[75] Although Buddhist education is said to have been 'visibly weakened as the state began to take over educational responsibilities',[76] the pedagogical approach practised especially in monasteries of the Ch'an tendency, stressing the importance of role models, school regulations and the ritualization of simple tasks in daily life, may still have influenced the organization of other schools outside the state system, which underwent a significant expansion under the Southern Song.[77] The truly novel development in Song times was the proliferation of local schools or 'academies'. Between at least 56 and 73 of these academies were founded under the Northern Song, and another 260–317 under the Southern Song. Academies could grow from pre-existing clan or family schools or emerge as new creations under the aegis of, for instance, a reputed scholar or a government official. While the organization of these schools may have been inspired by Buddhist examples, it was in particular the rising movement of Neo-Confucian reformers under the Southern Song that promoted their spread throughout the empire.[78]

---

[73] Lee, 'Sung education', 107–108, 113, Meskill, *Academies*, 13.

[74] Lee, 'Sung education', 107–108.

[75] Kuhn, *Age of Confucian rule*, 109, Meskill, *Academies*, 8–11, Chün-fang Yü, 'Ch'an education', 61–98.

[76] Lee, 'Sung education', 135.

[77] Meskill, *Academies*, 9–10, Chün-fang Yü, 'Ch'an education', 99–100, Lee, *Education*, 84–85.

[78] Meskill, *Academies*, 11–16, Elman, *Cultural history*, 58, 651, Lee, 'Sung education', 112–113, 135–136, De Bary, 'Chu Hsi's aims as an educator', 202, 210–211, 216–217, Wing-tsit Cha, 'Chu Hsi and the academies', *passim*, Friedsam, 'L'enseignement', 55, Kuhn, *Age of Confucian rule*, 126.

The Yuan sponsored both government schools and local initiatives. Local government schools were supported by land grants. Nearly all of these schools also had Confucian temples, and often Yuan local schools were even known as 'temple schools'. Moreover, the Yuan were the first to envisage the founding of 'community schools': schools where 'locally respected scholars' would teach young children during periods when farming activities had ceased. The spread of academies that would prepare students for civil service examinations was heartily encouraged, too.[79]

The first Ming emperors expanded the state school system with a vengeance. The overriding aim of their educational policy was to impregnate the population of the empire, after the 'unfortunate' interlude of 'foreign' domination under the Yuan dynasty, with state-sanctioned Confucian values. In 1369, the Hongwu emperor ordered that every prefecture, subprefecture and county should establish a school. Together with the Imperial University in the capital, these state schools were initially intended as the only authorized gateway to the civil service examinations.[80] A community school system, established by Hongwu in 1375, aimed to teach Confucian rites and rituals especially to children in rural communities. In practice, these schools sometimes also helped young boys to prepare for civil service examinations.[81] The few local academies that had remained in existence after the upheavals at the end of the Yuan languished during the early Ming because of the palpable official preference for the newly-founded state schools.[82]

Yet, after a few decades, the official school system once again failed to meet the growing demand for education from below, due to the lack of funds, teachers and consistent support from the imperial authorities.[83] Educational facilities outside the state school system began to show renewed expansion from the mid-fifteenth century onwards. This time, Buddhist or other religious institutions no longer played as significant a role in the schooling of the common population as they had done in the past, although Buddhist temple schools continued to exist in Ming and Qing times.[84] Aside from family or lineage schools, it was now local academies in particular that supplied the lion's share of educational provision

[79] Lee, *Education*, 91–94.
[80] Lee, *Education*, 96–97, Brook, 'Edifying knowledge', 105, Meskill, *Academies*, 21–22.
[81] Leung, 'Elementary education', 382, Lee, *Education*, 98–99.
[82] Meskill, *Academies*, 21–22, Lee, *Education*, 99–100.
[83] Meskill, *Academies*, 22–25, Leung, 'Elementary education', 382–383.
[84] Elman, *Cultural history*, 246.

outside the official system. The rate at which academies were established or restored increased from an average of 0.9 a year before 1470, via 3.43 a year between 1470 and 1505, to 7.5 a year between 1506 and 1572. Academies offered both preparatory education for civil service examinations and an opportunity to cultivate learning for its own sake. They were funded from a mixture of private and public sources, including income from rents, donations and possibly contributions from wealthy merchants.[85] As academies spread throughout the country and started to generate unorthodox and reformist ideas, powerful factions in the imperial bureaucracy increasingly perceived them as hotbeds of political discord and eventually persuaded the emperor to clamp down on this 'threatening' movement. Although the edict of 1579 ordering the abolition of all private academies was not implemented to the letter, it *did* lead to a marked slowdown in their growth for the remainder of the Ming regime. The annual index of new academies founded in the period between 1573 and 1620 fell to 4.7.[86]

Under the Qing, the system of formal education expanded on even grander scale, although it fell short of a campaign of universal education.[87] In addition to an enlarged complex of educational institutions in the capital and a broad network of state schools at prefectural and county levels, which prepared candidates for the civil service examinations, a vast array of 'charitable' or community schools was created by imperial magistrates and local leaders, which were specifically aimed at providing elementary education to the children of the poor. Family or lineage schools, meanwhile, continued to offer elementary education for members of their own groups.[88] Local academies became much more numerous than under the previous dynasty. While there was an estimated 1,200 academies during the Ming, their number soared to 1,900 under the Qing. During the latter regime, they proliferated not merely in their traditional heartland, the Lower Yangzi area, but also in more distant regions of the empire, such as Yunnan, Gansu and Xinjiang. In contrast with Ming times, only a minority of these academies (probably some 10%) operated entirely beyond the influence of the state.[89] Local academies provided an alternative route for higher education next to the state school system. Moreover, they created a

85 Meskill, *Academies*, 25, 62–66.

86 Elman, *Cultural history*, 128, 131, 208–210, 651, Meskill, *Academies*, 87–139, Lee, *Education*, 100–103.

87 Woodside and Elman, 'Afterword', 528–529.

88 Leung, 'Elementary education', 384–391, Elman, 'Civil service examinations', 113, 116, Woodside and Elman, 'Afterword', 526–528, Elman, *Cultural history*, 246, 376.

89 Woodside and Elman, 'Afterword', 526–527.

fertile environment for the flowering of so-called 'evidential' scholarship; that is, research aimed at recovering and restoring past Confucian learning by the rigorous, critical study of classical sources.[90]

Formal education in China from the Song onwards thus tended increasingly towards a system composed of a multi-layered state school complex and a network of private schools, which to some extent were interconnected and shared a Neo-Confucian ideology. Religious institutions were increasingly relegated to a subordinate position. In Europe, by contrast, religious institutions were not only initially much more significant for the supply of formal education than state governments, but they continued, or resumed, their prominent role as providers of educational facilities even after secular institutions had distinctly come to the fore. In the very period that the Tang created a empire-wide state school system in China, monasteries and cathedrals were providing the basic educational facilities in Europe, except for a handful of court or palace schools which offered a measure of education to scions of the higher aristocracy.

At some times and in some places, the creation of monastic and episcopal schools was admittedly promoted by territorial rulers. In 789, for example, Emperor Charlemagne ordered that every monastery and every diocese should provide instruction in singing, calendar calculation and grammar and that every school should have a number of books, although the implementation of this order surely must have fallen short of its ambitious aims.[91] But the chief impetus for the establishment of these schools connected to religious institutions came from the clergy themselves, who sought an effective means to educate new candidates for their ranks.[92] Monastic schools, located in monasteries or convents, and episcopal schools, run by a chapter of canons attached to a specific church (often a cathedral), were first and foremost intended to train future clergymen. For several reasons, these institutions nevertheless exerted a far-reaching influence on formal education in society at large. First, because of the sheer growth of the ecclesiastical establishment itself: the estimated number of monasteries alone in Latin Christendom rose from 1,100 in the early sixth century, via more than 4,000 in the ninth century and 12,500 in the

[90] Elman, *From philosophy to philology*, 7–36, 79–80, 105–106, Woodside, 'Divorce', 476–485.

[91] Wriedt, 'Continuity and competition', 172–173.

[92] Graff, *Legacies of literacy*, 43–50, Wriedt, 'Continuity and competition', 172–173, Orme, *Medieval schools*, 17–27.

eleventh century, to a peak of nearly 22,000 in the twelfth century.[93] This implied, of course, that the mass of literate people in absolute and relative terms also substantially increased. Second, at least until the twelfth century, monastic and episcopal schools were often also open to children of the laity and, in this manner, made a sizeable contribution to the education of groups in society outside the clergy. And third, monasteries in Europe, like those in Tang China, came to play a crucial role in the production and distribution of texts. It was at the Abbey of Saint-Martin at Tours where, in about 800, a simplified script and a standardized form of punctuation were developed that considerably eased the reproduction and reading of written texts.[94] Moreover, of all books produced in Europe up to the twelfth century, the vast majority came from monasteries.[95]

After about 1100, the pattern of formal education in the world of Latin Christendom converged with that in China for a while. There was a tendency for the relative weight of religious institutions, especially in the urban belt running from Northern Italy via Southern Germany and the Rhineland to the Low Countries, to decline. The sweeping monastic reform movement that gathered pace in Europe from the eleventh century onwards led many monasteries to abandon the time-honoured practice of admitting very young children as potential novices and often induced these communities to withdraw from elementary and secondary education altogether. Connections between monasteries and schooling were not severed everywhere, however. Whereas monasteries in Italy after 1100 no longer provided elementary or secondary education, monastic schools in England, for example, continued to exist until the Reformation in the late 1530s. For centuries, Benedictine and Cistercian monasteries and nunneries in England persisted in providing education both to their own novices and to lay boys and girls.[96] Moreover, from the thirteenth and fourteenth centuries across Europe, religious houses of newly-founded mendicant orders (such as the Franciscans, the Dominicans or Augustinians) and communities of orders of regular canons frequently offered instruction to their own members (and sometimes, to a few outsiders as well) in grammar, logic, philosophy and theology, in preparation for further study at a university.[97] Another new religious association, the Brethren of the

93 Buringh and Van Zanden, 'Charting the "Rise of the West"', 427.
94 Blum and Dudley, 'Standardized Latin', esp. 224–225.
95 Buringh and Van Zanden, 'Charting the "Rise of the West"', 426–429.
96 Grendler, *Schooling*, 6–7, Orme, *Medieval schools*, 255–276, 287.
97 Grendler, *Schooling*, 8, Orme, *Medieval schools*, 261–265, 287.

Common Life, which flourished especially in the Low Countries from the fifteenth century onwards, founded numerous boarding schools and Latin schools which provided multi-graded instruction both in Latin and the arts.[98] Chapter or cathedral schools, meanwhile, underwent a decline in Late Medieval Italy, but continued to exist in other parts of Europe. In some regions, such as England, a small number of parish schools arose, which offered instruction at an elementary level to children from the local community. At the highest level of education, ecclesiastical authorities played a pivotal role in the establishment and organization of the first universities in Italy, France, England and other countries. In the early stages, Dominicans and Franciscans often formed a large proportion of the teaching staff.[99]

Yet, from the twelfth century onwards, the weight of religious institutions tended to decline in relative terms because public and private educational provisions expanded much more quickly and massively. In many cities, starting in North Italy in the thirteenth century and also later in lands across the Alps, schools for elementary and secondary education were created that were not under the control of the Church, but that were run by civic authorities. Besides instruction in basic reading skills, these urban schools offered either a curriculum with an emphasis on Latin grammar, oratory and writing, supplemented with a smattering of vernacular reading, or a curriculum in the vernacular focused on the acquisition of computational skills (*abbaco*).

The driving force of this remarkable expansion of a 'secular', city-based schooling system in the High and Late Middle Ages was a combination of policy initiatives by urban governments and increasing demand for education from broad sections of the urban population, ranging from nobles and merchants to artisans and shopkeepers (and in some cases, even the labouring class). For various reasons, both urban inhabitants and urban magistrates found these new facilities eminently useful. Grammar or 'Latin' schools and reckoning or 'abbaco' schools offered people more opportunities to obtain the requisite knowledge and skills for careers as merchants, professionals or administrators than monastic or episcopal schools could supply, and they provided towns as a whole with a growing pool of qualified personnel and a body of citizens imbued with an

---

[98] De Ridder-Symoens, 'Sécularisation', 729–730.

[99] Graff, *Legacies of literacy*, 57–59, Grendler, *Schooling*, 8–10, Orme, *Medieval schools*, 62–66, 39–195, 244, 253, De Ridder-Symoens, 'Sécularisation', 726–727, 733.

ethos of civic virtue.[100] The demand for advanced skills, especially in leading centres of trade and industry such as Florence, Venice, Nuremberg or Antwerp, grew to the extent that in the Late Middle Ages, these cities also saw the rise of sundry independent schoolmasters – i.e. masters not funded from the city purse, though often operating with the authorization of urban magistrates – who made a living by giving, for a fee, instruction in mathematics, Latin or some other subject in a boarding school or at home.[101]

Secular powers also gradually gained more influence over institutes for higher learning. Universities multiplied across Europe after 1100. Starting from a mere six in the twelfth century (four in Italy and one each in France and England), their numbers rose via twenty in the thirteenth century and 44 in the fourteenth century to 70 in the late fifteenth century, of which nine were located in Spain, fifteen in France, seventeen in Italy and more than twenty in Central Europe. By then, universities had largely come under the sway of territorial rulers and urban governments, who saw these institutes as valuable breeding grounds for well-trained administrators and professionals.[102]

Monasteries, meanwhile, did not only lose relative importance as hubs of education but also as centres for the production and distribution of texts. In the very period when demand for manuscript books soared, production expanded and the production process quickened thanks to the introduction of simplified and mechanized binding techniques, the sites of production increasingly moved outside the walls of the cloisters. The production of manuscripts increasingly came under the control of entrepreneurial booksellers in cities, especially university towns, who subcontracted the different operations to groups of workers in a kind of cottage industry. The actual task of copying was often carried out by lay scribes at home instead of by monks in *scriptoria*.[103]

From the sixteenth century onwards, the distance between China and Europe widened again as religious institutions in Europe, in contrast with China, made a remarkable comeback in elementary, secondary and higher

---

[100] Grendler, *Schooling*, 11–23, 29–34, 87–141, 306–318, De Ridder-Symoens, 'Sécularisation', 726–729, Kagan, *Students and society*, 10–11,17–18, 42–43, Wriedt, 'Continuity and competition', 207, Neugenauer, 'Niedere Schulen', 215–216.

[101] Grendler, *Schooling*, 29–34, De Ridder-Symoens, 'Sécularisation', 729–730, Swertz, *Capitalism and mathematics*, 15–17.

[102] De Ridder-Symoens, 'Sécularisation', 733–734, Grendler, *Schooling*, 23–29, Buringh and Van Zanden. 'Charting the "Rise of the West"', 429–431.

[103] Petrucci, *Writers and readers*, 172–173.

education. Differences within Europe increased, too. The twin movements of the Protestant and Catholic Reformation in the sixteenth and seventeenth centuries produced different effects. However, in contrast with what 'Weberians' assume, these different effects can *not* be found in disparities in human formation. Levels of literacy or numeracy in Catholic regions in the early modern period did *not* persistently lag behind those in Protestant regions, as we have seen. Rather, the main differences between Protestant and Catholic regions resided in the extent and manner in which religious institutions were involved in education.

In regions in Europe that took the side of the Protestant Reformation, monasteries and convents were dissolved and the property of the Catholic Church was seized. This deathblow against the ecclesiastical establishment often meant that existing schooling facilities offered by religious institutions could not longer be maintained, at least not in a traditional way. But this did not spell doom for the supply of formal education as such. In most Protestant countries, urban magistrates or territorial rulers assumed the functions that had formerly been performed by monastic or cathedral schools, and often even extended the array of educational provision. Once a city or territory in the Holy Roman Empire had embraced the Lutheran confession, the municipal authorities or the territorial prince usually issued a *Schulordnung* which required religious instruction for the young. For this purpose, existing schools in cities were transformed into vehicles for popular education in the Lutheran spirit and new schools were established in parishes in the countryside. Expanding educational facilities was not the same as increasing literacy, however. The emphasis in elementary instruction, at least in rural areas, was for a long time on the memorization and recitation of catechisms rather than on reading or writing skills. It was not until after 1700 that enhancing literacy became an important goal in education, as a prerequisite for the ability of laymen to read the Bible.[104] In addition to the expansion of popular schooling, the infrastructure of higher learning was also developed. New universities arose to train ministers for Lutheran churches. Most of the funding for these educational facilities was supplied by revenues from the former properties of monasteries and other establishments of the Catholic Church. In the territory of Hesse-Kassel between c.1520 and 1590, for instance, 38% of the revenues from assets of convents and monasteries seized by the

[104] Van Horn Melton, *Compulsory schooling*, 4–5, 8, 38–39, Neugebauer, 'Niedere Schulen', 219–224.

state were used for administrative purposes, 3% for rewards to faithful servants and 59% for expenditure on education and the maintenance of Lutheran clergy, parishes and hospitals. A substantial share of the latter item was earmarked for the upkeep of a new university in Marburg, which was envisaged to serve as a bulwark of the Protestant religion.[105]

Switzerland, the Dutch Republic and England, which each in their own way espoused the cause of the Protestant Reformation, also saw momentous changes in the field of education. The magistrate in Zwinglian Zürich, for example, took over control of the Latin schools from the formerly Catholic churches, extended their supervision over private, vernacular schools, established an academy for higher education and introduced a system of scholarships and bursaries to enable citizens' sons to study abroad – largely paid for from the revenues of former canons of the church.[106] In the Calvinist Dutch Republic, properties of religious houses that after the dissolution of the monasteries during the Revolt were not reclaimed by the nobility, but came under the administration of the provincial states. As in Hesse-Kassel, the revenues of these ecclesiastical properties were mainly allocated *ad pios usos*, a purpose that was admittedly conceived in a very broad way. Apart from generating means to pay pensions to former monks and nuns, salaries to Reformed ministers and wages to village schoolmasters, and to defray outlay on newly-founded universities (Leiden, Groningen and Franeker), which, like Marburg, were first and foremost intended as training centres for Protestant clergy, the erstwhile monastic properties also served as sources of funding for such purposes as financing the war against Spain or keeping the West-India Company afloat, which were also said to be essential for the defence of the True Religion.[107]

In England, revenues from the dissolution of the monasteries and the seizure of church property in the 1530s and 1540s were not directly used for pious or educational purposes.[108] More than two-thirds of the monastic lands seized by the Tudor authorities had in fact already changed hands by 1547. The vast majority of the new owners consisted of established local

105 Wolff, *Die Säkularisierung*, 13, 132–161, 368–384, Seifert, 'Das höhere Schulwesen', 284–292.

106 Maag, 'Financing education', 205–207.

107 Van Beeck Calkoen, *Onderzoek naar den rechtstoestand*, 35–71, 209–238, Van Nierop, *Van ridders tot regenten*, 139–141 Maris, *Reformatie der geestelijke en kerkelijke goederen in Gelderland*, 33–39, 42–48, 56–69, Van Apeldoorn, *De kerkelijke goederen in Friesland*, vol. II, 369–431, Roelfsema, *De klooster- en proosdijgoederen in de provincie Groningen*, 72–79.

108 Carter, 'Fiscal Reformation', 96–97.

families, mostly gentry stock.[109] The 'educational revolution' detected by Lawrence Stone, consisting of an upsurge in literacy and a growing participation in higher education between 1540 and 1640, was not a direct consequence of the Reformation. If there was a connection, it was rather an indirect one: social groups which had benefited from the massive transfer of former ecclesiastical property were prepared to invest more in higher education by giving lavish donations to university colleges and teaching staff and by increasing charitable gifts to poorer students.[110]

In regions in Europe where Catholicism retained, or regained, the upper hand, religious institutions gained *more*, not *less*, control over education in the early modern period. On the whole, the supply of educational provisions in these areas expanded rather than contracted. In the sixteenth century, Hapsburg Spain went through an 'educational revolution' of its own. Primary education expanded, literacy rates went up, secondary schools multiplied and the number of universities rose from six in 1474 to 33 in 1620. This extension in facilities for formal learning was initially largely due to the driving force of municipal authorities, who saw the acquisition of reading and writing skills and an acquaintance with humanist culture as eminently desirable for the formation of good citizens, and of the Spanish Kings, who were particularly interested in using universities for training clergymen and state officials.[111] In the diocese of Cuenca, two-thirds of the literate defendants questioned by the Inquisition between 1540 and 1661 declared that they had learned their skills from a public schoolmaster, against only 14% from a village priest or sexton and 16% from a family member or friend.[112] From the late sixteenth century, however, the share of religious institutions in the supply of education was on the increase again. Due to a shortage of money and a growing concern for the social and economic disadvantages of 'excessive' education, municipal authorities became less able and less willing to continue to support elementary and secondary schools, and after the 1620s even found themselves compelled by the state to cut the number of Latin schools. Religious orders, by contrast, considerably extended their network of facilities for elementary and particularly secondary education.[113]

---

109 Hoskins, *Age of plunder*, 135–138, Overton, *Agricultural revolution*, 168–169.
110 Stone, 'Educational revolution', 67–68.
111 Kagan, *Students and society*, 18–20, 35–36, 41–42, 63–64, 70.
112 Nalle, 'Literacy', 75.
113 Kagan, *Students and society*, 21, 44–46.

In particular, the Society of Jesus, at the behest of local magistrates in several cities, took over control of existing municipal schools and, beginning in Gandía (near Valencia) in 1547, founded numerous new colleges of its own. By 1700, there were no fewer than 118 Jesuit colleges all over Spain. Jesuit colleges offered instruction free of charge at various levels of education, ranging from introducing students to reading and writing to university-level teaching. While on the one hand, the rise of Jesuit colleges to some extent compensated for the collapse of public primary schools, on the other hand, it deprived universities of potential students. With regard to secondary education, the Society of Jesus, according to Richard Kagan, 'quickly became the leading organizer' in Hapsburg Spain. Other religious orders followed the Jesuits' trail. Piarists opened their first elementary schools in the 1670s, although their network did not spread very widely until the second half of the eighteenth century. Franciscans, Dominicans and the Augustinians became active in secondary schooling after 1700. The share of regular clergy among students and teachers at universities from the seventeenth century onwards also increased significantly. Although municipal schooling and instruction by private teachers showed a marked revival after 1714 under the Bourbon regime, secondary schools led by religious orders in the 1760s still claimed nearly one-third of all pupils. Eighty per cent of these attended Jesuit colleges. The expulsion of the Jesuits a few years later was a blow from which the Spanish education system did not easily recover.[114] Furthermore, from the early seventeenth century, religious orders provided the lion's share of secondary education in the Spanish Netherlands. Between 1590 and 1630, Jesuits, Augustinians and others opened nearly 70 new colleges.[115]

In Italy, the Catholic Reformation after 1550 led to a massive growth of catechism schools under the charge of lay confraternities, which offered free vernacular instruction on Sundays and holidays in reading, writing and religious doctrine to boys and girls from poor families. *Scuole pie*, founded by the Piarists from the 1590s onwards, gave boys from the lower classes full-time, free instruction in vernacular reading, writing, arithmetic and some Latin.[116] Jesuits, Piarists and other new religious orders, such as the Barnabists and the Somaschi, were called upon by municipal authorities in numerous places in Italy to take over existing communal

[114] Kagan, *Students and society*, 21–23, 50–58, 186–190, Viñao Frago, 'Alfabetización', 50–51.

[115] Leyder, 'L'éclosion scolaire', *passim*.

[116] Grendler, *Schooling*, 333–358, 381–386.

schools.[117] Moreover, they founded many new schools for secondary education, which catered in particular for boys from the middle and upper layers of society. Starting with a college in Messina in 1548, the number of Jesuit colleges alone rose from eighteen in 1556 via 30 in 1565, 49 in 1600, and 80 in 1630 to 111 at the end of the seventeenth century. Some of these colleges, like the *Collegio Romano* in Rome, functioned in fact as institutes of higher learning or offered instruction at a level equal to that provided by universities.[118]

A similar extension of the role of religious institutions in education took place in France. The driving force for the seventeenth-century expansion in primary education in rural areas was the Catholic Church. It was the bishops who persuaded the royal government in 1698 to enact a regulation to the effect that each parish in France had to be provided with a school that offered, at a minimum, instruction in reading and religious doctrine. When, in the eighteenth century, state officials tried to curb the spread of popular education in the countryside, it was the Church that 'vividly resisted this push against the schools from above'.[119] As in Spain and Italy, in many towns in France, Jesuits and other new religious orders, such as the Oratorians and the Doctrinaires, were requested to assume control over existing municipal schools. In addition, they established a vast number of new schools, colleges and other institutes for secondary or higher education. The number of Jesuit colleges rose from two in 1560 to 34 in 1600, 94 in 1640 and more than 100 in 1700. Oratorians founded more than twenty colleges in the course of the seventeenth century, the Doctrinaires nearly 30.[120]

From the seventeenth century onwards, in the Catholic territories of the Holy Roman Empire, Latin schools (*Gymnasien*) and other institutes of higher education largely came into the hands of the Society of Jesus, which, as a matter of principle, offered instruction free of charge. More than 50 Jesuit *Gymnasien* were founded in the first half of the seventeenth century alone. The universities of Mainz, Würzburg, Ingolstadt and Graz became veritable hubs of Jesuit learning.[121] In places where secondary education was not dominated by Jesuits, schooling was usually provided

---

[117] Grendler, *Schooling*, 396–396.

[118] Grendler, *Schooling*, 370–371, Borgato, 'Mathematical research', 131.

[119] Chartier c.s, *L'Éducation*, 5–16, 37–41.

[120] Chartier c.s, *L'Éducation*, 61, 78–80, 165–168, 186–188, Graff, *Legacies of literacy*, 148–149, Huppert, *Public schools*.

[121] Hammerstein and Müller, 'Das katholische Gymnasialwesen', 324–336, Seifert, 'Das höhere Schulwesen', 320–328, Hellyer, *Catholic physics*, *passim*.

by Piarists, Benedictines or mendicant orders.[122] While from the late sixteenth century, parish schooling in rural areas also expanded, for a long time, the emphasis remained on singing and oral and visual religious instruction rather than reading, writing or arithmetic. Images and sounds were more important than words and figures. After 1770, however, the system of elementary education in Austria was fundamentally reformed by the Hapsburg authorities, with broad support from the ecclesiastical hierarchy. The *Allgemeine Schulordnung* issued by Empress Maria Theresia in 1774 stipulated that henceforth every child between the ages of six and thirteen should learn to read and write. The expansion of the school system that accompanied this declaration was, ironically, for a large part made possible by the seizure of properties of the Society of Jesus.[123] Austria's educational reforms were soon imitated in several other Catholic territories in Central Europe, and nowhere more eagerly than in principalities ruled by bishops or abbots.[124]

Variations in Vocational and Technical Education

The growing divergence in formal educational facilities between China and Europe, and within Europe itself after c.1500, also manifested itself in the area of vocational and technical education. This divergence is a further reflection of disparities in perceptions of human capital formation. It suggests that at certain times and in some places, there was an increasing need not only to establish formal requirements for particular tasks or vocations, but also to create formal organizational structures in which the required knowledge and skills could be obtained. This need could arise both from 'above' (for example, from changes in government policies or in market capitalism) or from 'below', that is, from individuals or groups who aspired to improve their economic position or to enhance their status or power in society.

In China, schools with a vocational emphasis were few and far between before the middle of the nineteenth century. Little evidence has come to light for Song, Ming or early Qing times that documents the existence of schools for vocational education, and for technical education in particular. No traces have as yet been found of educational facilities which might have offered regular courses on such topics as bookkeeping, surveying,

122 Hammerstein and Müller, 'Das katholische Gymnasialwesen', 336–343.
123 Van Horn Melton, *Compulsory schooling*, 6–9, 61, 76–80, 232–233.
124 Neugebauer, 'Niedere Schulen', 241–243.

navigation or engineering. The technical training of hydraulic experts, for example, was approached in a purely ad hoc manner. Hydraulic engineers acquired their knowledge 'on the job from subordinates or from the writings of their predecessors'.[125] Mining schools did not exist.[126]

Nevertheless, the dearth of evidence about vocational or technical training does not exclude the possibility that in some places, some kind of instruction on technical subjects in a tutorial relationship or classroom setting actually may have taken place before the mid-nineteenth century. The pervasive influence of the civil service examination system and the scholar-official ideal on Chinese education did not entirely preclude the emergence and persistence of bodies of knowledge outside the mainstream of scholarship. Specialist schools for law, mathematics, medicine and the military arts had existed in the imperial capital since the Tang or Song times, and some private academies may have taught mathematics and law, too.[127] Under the early Qing, officers' schools were created in Beijing to train Manchu banner men in military arts, translation, mathematics and astronomy.[128] Outside the capital, by the sixteenth century, the growth of commerce led to the appearance of arithmetic manuals for merchants that addressed all sorts of accounting problems. The heart of this commercial arithmetic tradition, which was also transmitted in print, presumably lay in the Lower Yangzi region.[129] In some places, private teachers of mathematics were also active, although such 'masters' were said to be 'not easy to find'.[130] With regard to religious institutions, however, it can safely be concluded that Buddhist and Daoist temples and monasteries did not offer any kind of vocational training, except for the education of their own religious personnel.[131] On the whole, technical training in China before the nineteenth century overwhelmingly took place via informal learning in the framework of families, partly aided by the use of books.[132]

---

[125] Dodgen, *Controlling the Dragon*, 7–8, 22.

[126] Vogel, 'Mining industry', 174.

[127] McMullen, *State and scholars*, 18, Friedsam, 'L'enseignement', 49–56, Lee, *Education*, 517–519, Kuhn, *Age of Confucian rule*, 126–127.

[128] Crossley, 'Manchu education', 356, 359, Jami, 'Learning mathematical sciences', 237.

[129] Friedsam, 'L'enseignement', 67–68, Gardella, 'Squaring accounts', 322–323, Jami, 'Learning mathematical sciences', 230.

[130] Jami, 'Learning mathematical sciences', 243–244.

[131] See e.g. Krohn, *Monastic life*, 66.

[132] Lee, *Education*, 523–541.

In Europe, by contrast, formal learning branched out into vocational schooling well before the nineteenth century. Starting with *abbaco* schools in Renaissance Italy, which offered basic accounting skills for merchants, a diverse complex of facilities for vocational and technical education came into being in the early modern period. These facilities ranged from curricular teaching on particular areas or topics to more general vocational training in a classroom context. Vocational education broadened in the sixteenth century with the creation of provisions specifically aimed at training highly-skilled craftsmen like navigators, engineers or surveyors. A further wave of extension took place in the eighteenth century. Within Europe, however, there were striking variations in the ways in which this complex of facilities evolved. These variations reflected different combinations between the principal forces that influenced the development of formal learning, namely governments, markets, religious institutions and demand 'from below'. Geographically, the variations largely, though not entirely, coincided with variations in dominant denominations. Catholic Spain, Italy and France underwent a different evolution from that which took place in Protestant countries in Northwest Europe, but also diverged from both Protestant and Catholic territories in Central Europe.

Spain was the forerunner in the sixteenth-century expansion of vocational education. The driving forces were agencies of the Castilian state. In around 1500, the *Casa de la Contratación* (House of Trade) in Seville was the first organization in Europe to create institutional facilities for the training of navigators. The task of instructing pilots bound for the Indies was entrusted to the principal cosmographer of the *Casa* and formal rules were drawn up for their examination. Classroom instruction and special attention to the principles of astronomical navigation were explicitly intended 'as a way of shortening the years of apprenticeship necessary to learn the craft of piloting'.[133] In 1583, the architect of San Lorenzo de El Escorial, Juan de Herrera, masterminded the foundation of another, even more ambitious, institution for technical education, the Royal Mathematical Academy in Madrid. The purpose of the Academy was to train young people for professions based on mathematical knowledge and skills, such as engineers, gunners, instrument-makers, navigators, astrologers, architects, surveyors, designers of fortifications, builders of waterworks, painters, mechanics and experts in perspective and optics.[134] The Royal

133 Portuondo, *Secret science*, 57, 96.

134 Portuondo, *Secret science*, 82–84, Navarro-Brotóns, 'Teaching', 212–216.

Mathematical Academy remained in existence as a separate institute until 1625, while the educational facilities at the *Casa de la Contratación* continued to function until the transference of the *Casa* from Seville to Cádiz in 1717.[135] Further state-sponsored initiatives for the training of navigators included the foundation of the *Colegio de San Telmo* in Seville in 1681, the *Academia de Guardias Marinas* in Cádiz in 1717 and naval schools in Cádiz, Cartagena and El Ferrol in the mid-eighteenth century. Private schools for navigation did not arise in great numbers until after c.1770.[136]

In the sixteenth century, neither the old nor the new religious orders were able to assist the Spanish state in its programmes of vocational training, because they still lacked the necessary mathematical expertise. Even in the Society of Jesus, accomplished mathematicians were at that time hard to find.[137] However, the fact that the mission of the Royal Mathematical Academy in 1625 was incorporated into the *Colegio Imperial de San Isidro*, staffed by Jesuit teachers, reveals that the competence of religious institutions in vocational training had visibly increased by the second quarter of the seventeenth century.[138]

Religious organizations in this field also increasingly came to the fore in Italy and France. One of the first technical schools in Italy, the *Accademia del Disegno* in Florence, has been described as 'a kind of quasi-state institution'. Founded in 1563 by Giorgio Vasari for the purpose of training painters, sculptors and architects, the *Accademia* depended heavily on the patronage of the Dukes of the House of Medici and served as 'a consulting agency on all engineering projects and public works' of the Medici state.[139] The *scuole pie*, which arose in cities and towns all over Italy in the first half of the seventeenth century, were a creation of a religious organization, the Piarists. Although these schools did not train pupils for specific occupations, they *did* give poor boys a head start for careers as artisans or merchants by offering free education in vernacular reading, writing and *abbaco* (and some Latin).[140]

---

135 Goodman, *Power and penury*, 127–129, Arroyo Ruiz-Zorrilla, *Apunte para una historia de la enseñanza de la nautica*, 70–71.

136 García Garallón, 'Education of pilots', 191–196, Tikoff, 'Saint Elmo's orphans', 4–7, Lopez Piñero, *Ciencia*, 454–455.

137 Gatto, 'Christoph Clavius' "Ordo Servandus in Addiscendis Disciplinis Mathematicis" ', *passim*.

138 Portuondo, *Secret science*, 86.

139 Henninger-Voss, 'Measures of success', 151–152.

140 Grendler, *Schooling*, 381–391.

In France, as in Spain, it was the central government that initially led the way in creating facilities for vocational education. Colbert's grand design to endow France with a powerful navy involved, among other things, the foundation of royal chairs in navigation in Le Hâvre (1660) and Nantes (1672) and the establishment of training colleges for naval officers in Brest, Rochefort and Toulon (1682).[141] In contrast with Spain, however, from the very start, the realization of this educational scheme was made possible by new religious organizations that arose during the Catholic Reformation. Many teaching posts at the newly-created schools were filled by members of the Society of Jesus. Some of the posts were even established *in* Jesuit colleges, although the courses could also be attended by external pupils. Jesuits were very active in the spread of nautical knowledge in France. Indeed, the authors of the best-known French textbooks on the art of navigation published in the seventeenth century, Georges Fournier and Paul Hoste, were members of the Society of Jesus.[142]

While religious orders were not engaged in the staffing of higher institutions for the professional training of engineers that were established in the mid-eighteenth century, such as the *École des Ponts et Chaussées* (1747), the engineering school in Mezières (1748) or the *École des Mines* (1778), they *did* figure prominently in the military academies that were established by the government in 1776 to improve the quality of future army officers. In five of the twelve new academies, the teaching staff mainly consisted of monks belonging to the *Congrégation de Saint-Maur*. In 1788, no fewer than 46 Maurists were employed in the military academy of Sorèze alone.[143]

Religious institutions also became involved in technical education at an intermediate level.[144] The most active religious organization in this field was another product of the Catholic Reformation, the congregation of the Brothers of the Christian Schools that had been founded by Jean-Baptiste de la Salle in 1680. In the eighteenth century, the *Frères Chrétiennes* set up navigation schools in Vannes and Nantes and a school of commerce at Boulogne, offered courses in drawing and geometry in inland towns such as Cahors and Castres, and established boarding schools combining

141 Anthiaume, *Évolution et enseignement*, vol. II.

142 Russo, 'L' hydrographie', *passim*, Chartier c.s., *L'Éducation*, 228; in Spain, navigation was taught at the Jesuit college in Seville in the 1680s, see Arroyo Ruiz-Zorrilla, *Historia de la enseñanza de la nautica*, 70–71.

143 Chartier c.s., *L'Éducation*, 217–220, Artz, 'L'éducation technique', *passim*, McManners, *Church and society*, I, 600.

144 Brockliss, *French higher education*, 26–27.

instruction in applied sciences with vocational training for careers in surveying, accounting, navigation, horticulture and shop, office and government service in a dozen cities and towns in various parts of France. There was never a shortage of applicants. At the level of primary schooling, charity companies within parishes, founded under the impulse of the Catholic Reformation after 1650, and the *Frères Chrétiennes* provided free education in reading, writing, arithmetic and other skills for children from poor families, as preparation for their guild-supervised period of apprenticeship. Before the appearance of free drawing schools, created by painters under the patronage of local and royal government authorities after c.1750, these schools run by charities and the Christian Brothers completely dominated the field of pre-apprenticeship formal education in France.[145]

In Central Europe during the early modern period, for a long time, vocational education remained largely limited to an assortment of reckoning schools for teaching merchant skills. It was not until the eighteenth century that special institutions were created for the training of professionals such as army officers, army engineers, navigators, mining experts, architects and civil engineers. Academies for engineers and army officers were founded in Austria, Prussia, Saxony and Württemberg between c.1720 and 1770; navigation schools were erected in Hamburg (1749), Emden (1782), Bremen (1799) and several other port cities; *Bergakademien* and *Bergschulen* were established in Schemnitz (Hungary, c.1735), Freiberg (Saxony, 1765), Berlin (1770) and Clausthal (Hannover, 1775); and building schools appeared in Prussia and Baden in about 1800.[146] In Berlin in 1747, the educational reformer Johann Julius Hecker founded the first *Realschule*, a secondary school which offered instruction in a wide range of vocational subjects, such as bookkeeping, mining, engineering, printing and architecture, for sons of the middle and lower classes.[147] In the late eighteenth century, local initiatives in the numerous towns in Germany also led to the creation of drawing schools, building schools and 'schools of industry', where the children of the poor could learn to weave and spin.[148] In contrast with Italy or France, religious organizations were not involved in the creation or maintenance of any of these new facilities. The driving forces

---

[145] Prévot, *L'enseignement technique*, 15–19, 42–60, Chartier c.s., *L' Éducation*, 225–227, McManners, *Church and society*, I, 529–532, Crowston, 'From school to workshop', 46–62.

[146] Albrecht, 'Anfänge', 138–139, 141–144, Lundgreen, 'Engineering education', 41–46, Mehl, 'Anfänge', esp. 385–416, Brianta, 'Education', 278–281.

[147] Van Horn Melton, *Compulsory education*, 54–56, Albrecht, 'Anfänge', 131–136.

[148] Albrecht, 'Anfänge', 124–127, 139, 150, Bruchhäuser, 'Berufsbildung', 408.

were mostly state governments and, in some cases, urban governments or associations of private citizens.

Yet another pattern can be seen in Northwest Europe. In the Dutch Republic and England, neither religious institutions nor governments played an important part in the development of vocational schooling from the end of the sixteenth century onwards. Religious organizations in these countries were not active in creating or supporting provisions for technical education. With a few exceptions (such as the Reverend Petrus Plancius in Holland in the field of navigation in about 1600), Protestant clergy did not figure prominently among teachers or authors on technical subjects. In this regard, Anglican churchmen and Reformed ministers were no match for Jesuits, Maurists or Christian Brothers. In the Dutch Republic, for a long time, government agencies only supported technical education in an indirect way. Several institutes of higher learning, sponsored by provincial or urban governments, offered vernacular courses on mixed mathematics for a non-academic public. Just like the Royal Mathematical Academy in Madrid, the *Duytsche mathematicque*, founded as a branch of Leiden University in 1600, provided classroom instruction on surveying and fortification, which trained people for professions based on mathematical knowledge and skills. From 1600, the University of Franeker offered courses in Dutch on surveying, fortification and the art of navigation. The *Athenaeum Illustre* in Amsterdam, with a few interruptions, continued to provide vernacular instruction on the art of navigation from 1635 until 1838.[149] It was not until the mid-eighteenth century that public authorities began to contribute more directly to vocational education. Schools were founded for the training of army officers and military engineers. The Dutch Navy, jointly with the Dutch East India Company and local governments, established navigation schools in major port cities.[150]

In England prior to the eighteenth century, the main institution to offer some kind of vocational training was the Inns of Court. The London courts were the preferred destination for anyone who wanted to study common law.[151] In 1702, the state for the first time created a modest provision for training young people as naval officers by appointing schoolmasters to serve in operational warships. An academy for artillery officers and military engineers opened in 1741. In contrast with France or states

149 Davids, 'Universiteiten', 5–6, 8, 12.
150 Janssen, *Op weg naar Breda*, 53–62, 99–126, Davids, *Zeewezen en wetenschap*, 327–328.
151 Stone, 'Educational revolution', 51.

in Germany, however, England before the nineteenth century did not see the foundation of any state-sponsored institutions for the training of civil engineers or other groups of non-military technicians.[152]

However, the absence of religious institutions and the reticent attitude of governments did not imply that provisions for vocational training were lacking. Both in the Netherlands and England, facilities for technical education were amply supplied by private entrepreneurs, later assisted by non-profit institutions. Private schoolmasters in cities or in boarding schools in the countryside provided all kinds of courses on a commercial basis, preparing people for careers in trades such as surveying, gauging, bookkeeping, ship navigation or the design of waterworks and fortifications. Knowledge and skills for these trades were transmitted by classroom instruction or taught in tutorial settings. These 'private-venture teachers' in the Netherlands and England also made a mighty contribution to the fast-growing output of writings on mathematical subjects.[153] From the late seventeenth century onwards, vocational training by such entrepreneurs was extended to include parts of the manufacturing sector as well. Advertisements for courses on chemical subjects by private teachers, for example, began to appear in Amsterdam's leading newspaper, the *Amsterdamsche Courant*, from the 1680s onwards.[154] Later on, with the assistance of private, non-profit associations, this sort of vocational education also spread to the building trades. Those who aspired to become a master-mason or carpenter increasingly took to attending evening classes on drawing and mathematics – in addition to learning craft skills in a workshop by day – in order to improve their knowledge of the artistic and theoretical aspects of the building process. What was provided by governments and/or religious institutions in Spain, Italy, France or Central Europe was thus provided by market forces and non-profit institutions in the Netherlands and England.

Religious Institutions and Informal Learning

Formal learning was only part of the story. Throughout the pre-industrial period, informal learning remained essential for abilities that rested, to

[152] Lundgreen, 'Engineering education', 46.

[153] Davids, 'Ondernemers in kennis', 39, 41–44, Money, 'Teaching in the market place', 340–343.

[154] Van Nierop, 'Gegevens', 27 (1930), 271, 300, 28 (1931), 119.

a relatively large extent, on 'tacit' knowledge, such as skills in extracting, shaping or handling materials and material objects.[155] What role did religious institutions in China and Europe play in this informal route to the acquisition of knowledge and skills? The most effective way to answer that question is to identify settings in which connections between religious institutions and informal learning could occur, to find out what happened in those settings and to assess how important these settings were in the overall development of informal learning. This strategy may allow us to distinguish patterns of convergence or divergence between China and Europe and within Europe itself.

Before the twelfth century, the premier setting in which links were likely to be established between religious institutions and informal learning was that of the monasteries. In China and Latin Christendom in those days, monasteries were the religious establishments with by far the highest concentrations of land, labour and capital. As landowners, employers, investors, traders and consumers, monasteries weighed relatively heavy in the overall economy, too. The 4,600 Buddhist monasteries in China in the early 840s jointly owned an estimated 2% of all cultivated lands (with wide regional variations), which may have yielded more than 10% of the total agrarian produce of the Tang Empire.[156] The total number of people related to these monasteries and convents far exceeded the 260,500 fully ordained monks and nuns or the estimated 600,000–1,000,000 religious in the mid-ninth century, or the 400,000 monks and 61,000 nuns estimated for about 1220. Monks were not used to tilling the soil themselves; monastic lands were mostly worked by slaves and other categories of dependents, or they were leased to tenants.[157] Daoist monasteries, which emerged after the model of their Buddhist predecessors, albeit in smaller numbers, likewise possessed landed estates and employed outside labour to do much of the work.[158] Watermills formed a normal part of the capital assets of both Buddhist and Daoist monastic establishments. In addition, oil presses were sometimes part of the capital equipment.[159]

Overall estimates for the amount of land possessed by monasteries and convents in Europe are not available for the period before the sixteenth

---

155 Cf. Smith, *Body of the artisan*, 8.

156 Gernet, *Buddhism in Chinese society*, 12, 140.

157 Gernet, *Buddhism in Chinese society*, 12, 95–96, 115–116, 126–129, Ch'en, 'Economic background', 92–95, Kuhn, *Age of Confucian rule*, 109.

158 Kohn, *Monastic life*, 19, 35, 97, 101.

159 Kuhn, *Age of Confucian rule*, 109, Kohn, *Monastic life*, 97.

century, but this amount certainly increased as the number of monasteries grew apace from the Carolingian era onwards. Up to the twelfth century, monasteries were more often situated in rural areas than in towns. Much land was endowed to religious houses for the maintenance of the monks or nuns, in return for specific services (for example, the saying of prayers or the reading of masses) for the care of the souls of their donors. A substantial proportion of the monastic population was recruited from the nobility. In fact, this noble presence remained a constant element of the European monastic landscape until the end of the eighteenth century.[160] As in China, monks and nuns in Europe increasingly left the cultivation of the fields to dependents (lay brothers, tenants, wage labourers and the like); the construction of complex water systems, too, was mainly done by laymen. Monasteries in Europe were also important owners of capital goods, such as watermills or furnaces and forges. By the end of the tenth century, it is likely that most of the watermills in Western Europe were held by monastic houses. At that time, nearly all of these were corn mills, but mills for industrial purposes, such as fulling, were also added later.[161] Cistercian monasteries were especially active in the expansion of iron production after 1150.[162]

The claims advanced by Marc Bloch, Lewis Mumford, Lynn White and others to the effect that monastic orders played a pivotal role in land reclamation and technological innovation in early medieval Europe have been substantially qualified by research in recent years. In reality, monastic houses in Europe rarely undertook the arduous task of reclaiming waste areas themselves because they commonly received land that had already been brought into cultivation.[163] While monasteries often found themselves at the forefront of innovation in hydraulic technology with regard to the construction of complex water systems comprising aqueducts, pipes, fountains and the like,[164] they played a less conspicuous role in the diffusion of watermills. Most of the water-powered mills which

---

160 Beales, *Prosperity and plunder*, 4–7, Norborg, *Storföretaget Vadstena kloster*, 81–98; there were variations, of course. No monastic house in England, for example, as David Knowles pointed out, ever attempted 'to make itself into a preserve for the sons of noble or at least knightly families', Knowles, *Religious orders in England*, vol. II, 229–230.

161 Lawrence, *Medieval monasticism*, 178–180, Bouchard, *Holy entrepreneurs*, 96–97, 186–188, Lucas, *Wind, water, work*, 181, 196, Magnusson, *Water technology*, 54.

162 Fossier, 'Rural economy', 44.

163 Bouchard, *Holy entrepreneurs*, 97–106, 186–188, Mol, 'Aduard 1192–1594', 21, Van Bavel, *Goederenverwerving en goederenbeheer*, 270–275.

164 Magnusson, *Water technology*, 11–19.

monasteries possessed at the end of the tenth century had not been constructed by monks themselves. The use of water power was already fairly widely spread in Roman times and new watermills, including vertical-wheeled mills, were erected by numerous landowners across Western Europe in the Early Middle Ages.

What was new after c.800 was that ownership of mills increasingly moved from associations of 'share-holders' to individual owners. In northern Italy, France and England, a growing proportion of these owners consisted of ecclesiastical institutions, especially Benedictine monasteries. Paolo Squatriti has called this process of shifting ownership to monasteries by way of pious donations the 'clericalization' of water and milling.[165] Although some monastic orders, notably the Cistercians, introduced a few improvements in watermill technology and monasteries were slightly more advanced in the application of water power to industrial purposes, such as fulling or metalworking, than mill owners in general, the monastic sector on the whole did not stand out for its innovative behaviour in water milling.[166] Likewise, in the building or rebuilding of aqueducts and other facilities for water supply, rather than taking the lead, monastic houses followed the example of other ecclesiastical institutions.[167] Buddhist monasteries in China were not pioneers in watermill technology either. Water-powered mills had existed since the Han dynasty. During the Tang, watermills were not only owned and operated by monastic houses, but also by merchants and landed gentry. The Song government set up public watermills for grinding grain.[168]

The significance of monastic houses for economic and technological change in China and Europe between c.700 and 1100 consisted as much in their attitude to a growth in production and the spread of informal learning as in their role in the introduction and spread of new technology. Even though monasteries did not always behave like the Schumpeterian pioneers which generations of historians made them out to be, they *did* increase the area of arable land and they *did* help to enhance the total supply of agricultural produce available for consumption by society at large. Monasteries helped to expand agrarian production both directly,

165 Lucas, *Wind, water, work*, 160–161, 164, 176–178, Squatriti, *Water and society*, 125–145, esp. 144–145.

166 Böhm, 'Technische Bildung', 89–90, Lucas, *Wind, water, work*, 196.

167 Squatriti, *Water and society*, 20–23.

168 Needham, *Science and civilisation*, vol. IV, part 2, 370, 390–392, 400–401, Kuhn, *Age of Confucian rule*, 222–224.

by the exploitation of large farming estates, and indirectly, by letting out part of their land to tenants. Moreover, the very fact that much of the work in monastic houses was not carried out by monks or nuns themselves, but by labourers who did not form a integral part of the monastic community, implies that these religious institutions contributed more to informal learning than would have been the case if all work had been done in a 'closed shop'. As premier clusters of capital, land and labour that were open to a broad circle of people, monastic houses offered extraordinary opportunities for 'learning by doing' and 'learning by using' in a wide variety of occupations. Monasteries were an eminent site for acquiring and improving skills in constructing, maintaining and decorating a large building, tilling the land, growing crops, keeping animals, making iron, building and operating a watermill, copying and embellishing a text or, in China, printing a text by means of woodblocks. For centuries in China, Buddhist monks were the pre-eminent experts in building large, permanent bridges. From the Southern Song onwards, there are innumerable records of Buddhist monks being involved in the construction of bridges. In Buddhist writings, bridge-building was portrayed as an act of merit and charity.[169]

From the twelfth century, however, the relative importance of monasteries for informal learning began to diverge sharply, first between China and Europe, and later within Europe itself. While monks in China continued to enjoy tax-exempt status, from the Song onwards, they were increasingly *obliged* to assume responsibility for public works such as constructing roads or building bridges outside their own institutions. Monks were charged with performing all sorts of technical, organizational and financial tasks, such as purchasing the materials, recruiting the labour and providing the necessary funds for executing the projects, while barely receiving any compensation from the state.[170] From the Late Ming, monasteries were also forced to sell part of their estates and to pay taxes on the remaining landholdings as a means to help the imperial government to defray its growing military expenses. Local gentry frequently managed to take possession of large tracts of monastic land. In Southeast China, the share of monasteries in the total taxable property declined from more than 30% to less than five % by the end of the sixteenth century. As the upkeep of religious houses became ever more difficult to sustain, the

169 Kieschnick, *Impact of Buddhism*, 199–208.
170 T'ien Ju-K'ang. 'Decadence of Buddhist temples', 84–85, Dear, *Taoist ritual*, 32–33.

monastic population dwindled. Between the late fifteenth century and the 1660s, the number of Buddhist and Daoist monks and nuns dropped by more than 70%.[171] Meanwhile, the number of ordination certificates for Buddhist monks which were sold to line the government's purse soared to unprecedented heights. Eventually, the Ming regime even went so far as to distribute blank certificates without obliging purchasers to come to the capital to present themselves in person.[172]

As the Buddhist monastic order fell into prolonged decay,[173] the significance of monasteries as pools of technical knowledge and skills and settings for informal learning suffered too. In Fujian, under the early Qing, it was 'rare to see monks able to make contributions in the technical field', as they had done in Song times. In leading provincial monasteries, it was even 'very difficult to find a monk with adequate knowledge of bookkeeping to keep the temple's accounts'.[174]

In Europe on the eve of the Reformation, monasteries and convents still weighed heavily in the economy as both producers and owners of production factors. Guilds and town governments in the Burgundian-Hapsburg Netherlands, for example, complained in the fifteenth century that local crafts and trades suffered from competition from monastic houses of goods which – sometimes as subcontractors for urban entrepreneurs – were producing goods such as cloth, linen and beer for sale on the market.[175] In the Northern Netherlands, the share of monasteries in the total cultivated area in the early sixteenth century ranged from a few per cent in Holland to some 15% in Groningen and 20% in Friesland.[176] In England in the 1530s, some 20 to 30% of all the land in the kingdom was owned by the church. About half of the total income of these lands came from monasteries and convents, mostly in the form of rents received from manors.[177]

While the dissolution of monasteries and convents and the seizure of church property diminished the economic role of these institutions,

---

171 T'ien Ju-K'ang. 'Decadence of Buddhist temples', 92–96, Dear, *Taoist ritual*, 32–33, Davis, *Society and the supernatural*, 217.

172 Yü, *Renewal*, 155–162.

173 The external and internal causes of this decay, and contemporary attempts at reform, are discussed in Yü, *Renewal*, chapters 6,7 and 8.

174 T'ien Ju-K'ang, 'Decadence of Buddhist temples', 98.

175 Jongkees, *Staat en kerk*, 20, 91–103, De Man, 'Maatregelen', 284–286, Van Eeghen, *Vrouwenkloosters*, 40–42.

176 De Vries, *Dutch rural economy*, 41–42.

177 Hoskins, *The age of plunder*, 128–129, Overton, *Agricultural revolution in England*, 168.

including their contribution to informal learning, in England, the Northern Netherlands, Scandinavia and large parts of Germany, the story was quite different in Catholic regions of Europe. According to an estimate by Derek Beales, Catholic Europe in around 1750 had more than 25,000 monasteries and convents, which housed about 350,000 monks and nuns. Monastic establishments may have owned some 10% of the land, or about half of all the rural properties held by the Church.[178] Of all the Catholic states in Europe, perhaps the highest share has been recorded in the Electorate of Bavaria, where monastic houses in about 1800 controlled c.28% of all farms in the land.[179] Moreover, at that time, monastic houses still functioned as settings for informal learning. Bavarian monasteries in the early nineteenth century had, on average, 66 full-time and 96 part-time paid employees and also provided a kind of apprenticeship track. Apprentices formed some 25–30% of all craftsmen in monastic employment in the Electorate.[180]

Monasteries in China and Europe could be found in both towns and rural areas, often in isolated places like mountains, moors or marshes. After the eleventh century, towns became the centre of a building boom in religious structures. Cities in China and Europe came to be studded not only with monastic houses, but also with countless other places of worship in all sorts and sizes, ranging from mighty cathedrals and multi-storeyed pagodas, down to a multitude of smaller churches, temples, chapels and shrines. In Paris and its surrounding region alone, for instance, nearly 1,400 churches from the period between c.1140 and the middle of the thirteenth century have been preserved, all built in a style that has come to be known as 'Gothic'.[181] Except for synagogues, religious structures in Europe in this period were normally destined for Christian worship. While newly-erected temples in Song and Yuan China were sometimes affiliated to a particular 'Great Tradition', e.g. a Buddhist or a Daoist school or movement, they were often devoted to the cult of some local or supra-local deity, in which 'ritual masters' from different traditions could be involved. These temples were ordered in a kind of hierarchy. Peasants from several villages joined in worshiping societies that built a common temple, devoted to the local deity. Small, rural temples were in turn subordinated to a major temple in a nearby market town, which was more often than not

178 Beales, *Prosperity and plunder*, 2–4, 316.
179 Weiss, *Die Säkularisation der bayerischen Klöster*, 21.
180 Stutzer, *Klöster als Arbeitgeber um 1800*, 138–139, 141–142.
181 James, *Template makers*, 2.

erected by some rich landowner or merchant.[182] The building of religious structures and the building of towns were interrelated movements: growing towns erected new places of worship and the creation of new places of worship boosted the growth of towns. Richard von Glahn has drawn a direct parallel between what happened in China and Europe: 'Rising market towns of the Yangzi Delta bore a close resemblance to the cathedral towns of contemporary Europe. In both cases, the towns' affluence was inseparable from the magnificence of their religious monuments'.[183]

Were these booms in temple and church-building in China and Europe of equal importance for informal learning? The great age of temple-building in China did not last as long as the huge wave of church-building in Europe. The Chinese building boom by and large came to an end in the middle of the fourteenth century, when the empire was badly shaken by internal wars and the new dynasty that rose triumphantly from the fighting, the Ming, started to clamp down on all manifestations of religious life that did not conform with the newly-propagated cult of the state. As a consequence of this rigorous unifying policy, the number of Buddhist and Daoist temples and shrines, for instance, was dramatically reduced.[184] While by the end of the thirteenth century, the building drive in Europe had slowed down in its original heartland, Northern France, in other regions of Latin Christendom, such as Italy, Spain, the Holy Roman Empire, the Low Countries and England, vast numbers of cathedrals and other kinds of churches in the 'Gothic' style were erected in the fourteenth, fifteenth and early sixteenth centuries.[185] The churches that today dominate the skylines of Milan, Palma de Mallorca, Seville, Prague, Ulm, Utrecht, Antwerp and Ely, for example, were largely constructed in this very period.

Before the sixteenth century, building churches in Europe normally took much longer than erecting pagodas in China. On average, a pagoda was completed in less than ten years.[186] Smaller churches in Europe could be constructed in five or six years, but in most cases, the building process lasted for decades, or even for hundreds of years. Churches were usually built in a series of campaigns, alternated by periods of relative inactivity.

---

182 Von Glahn, 'Towns and temples', 189–195, Hansen, *Changing gods*, 9–10, 57, Ter Haar, 'Buddhist-inspired options', 95–98, 151, Davis, *Society and the supernatural*, 5–7.

183 Von Glahn, 'Towns and temples', 196.

184 Von Glahn, 'Towns and temples', 208–211.

185 Wilson, *Gothic cathedral*, chapter III, esp. 197, 224–231, 234–244, 280–282, Mark, 'Technological innovation', 16–17.

186 Guo, *Chinese architecture*, 64.

The background to this stop-go pattern was partly financial – the availability of funding – and partly technical. As churches in Europe in this period, in contrast with contemporary religious structures in China,[187] were largely built of stone and mortar instead of wood and brick, breaks were, according to John James, 'inevitable in even in the best-endowed works, for the nature of the lime mortar did not permit medieval craftsmen to complete the whole of a building in one operation'.[188]

Moreover, because of the nature of the materials used, building a church in Europe in the High and Late Middle Ages involved a greater variety of craftsmen than the construction of temples or pagodas in Song and Yuan China. Apart from carpenters, carvers or sculptors, it required the enlistment of masons, stonemasons, sawyers, glaziers, slaters, and, from the fourteenth century, often pipe organ builders as well.[189] Breaks during the building of a church did not mean that craftsmen had to remain idle. While some of them continued to be employed locally, others travelled to building sites elsewhere, where construction work continued. There was a continuous circulation of skilled labour within and between regions.[190] All these indications strongly suggest that between about 1100 and 1500, church-building sites in Europe offered opportunities for informal learning to a greater number and variety of people, and for a much longer time, than temple-building sites in contemporary China.

Informal learning took place at several levels. Architectural historians have drawn attention to the learning processes that went on between building sites. 'Structural lessons learned from previous building projects must have contributed to the success of new designs'. Builders learned from mistakes made in the past.[191] Informal learning also occurred on the sites themselves. Craft knowledge and skills in this period were generally transmitted orally and visually, and this also held true for crafts involved in the building industry. Sketchbooks or booklets on building techniques, such as those composed by master masons Villard de Honnecourt, Matthias Roriczer and Lorenz Lechler, were quite exceptional.[192] The ways in which masters applied geometrical techniques were often idiosyncratic

---

[187] Ledderose, *Ten thousand things*, 120–132.

[188] Guo, *Chinese architecture*, 57–73, James, *Template makers*, 2, 6.

[189] Binding, *Baubetrieb*, 285–334, Icher, *Oeuvriers*, 86–141.

[190] James, *Template makers*, *passim*, Binding, *Baubetrieb*, 184–185, Philipp, 'Eyn huys in manieren van eynre kirchen', 69–71, Böhm, 'Technik und Bildung', 104, Williams, *European organ*, 30, 97, 129.

[191] Mark, 'Technological innovation', 13–14.

[192] Shelby, 'Geometrical knowledge', 398, 406–420.

rather than standardized: every master devised his own rules of thumb.[193] It is precisely because so few things were put on record at the time that 'the vast bulk of the technical knowledge upon which mediaeval building and architecture were based disappeared with the dying of those oral traditions at the close of Gothic building in Europe', Lon Shelby has remarked.[194] It is for this very reason that reconstructing or replicating medieval building techniques still poses such a challenging task for historians. Learning thus almost exclusively took place on the job – by doing and by using. Some of the craftsmen in the building industry acquired their knowledge from family members, others learnt their trade from skilled workers in stone quarries or in the lodges or *Hütten* which served as the organizational framework for labourers on a building site. In most places, guilds did not become relevant supervisory institutions for the training of craftsmen until the Late Middle Ages.[195]

The schism in Latin Christendom in the sixteenth century led to a new building wave of religious structures, both in the Protestant and in the Catholic regions of Europe. However, new sacred buildings were much more rare in the former territories in the latter. Although Protestants in the early modern period *did* create a few monumental places for worship of their own, such as the *Westerkerk* in Amsterdam, St Paul's Cathedral in London or the *Frauenkirche* in Dresden, they usually restricted themselves to taking over existing churches or chapels and making them fit for Protestant worship by stripping them of most of their sacred objects. And because the Protestant Reformation wished to dispense with the idea of sacred spaces and places altogether, Protestants rarely erected the other sorts of religious buildings such as those found in Catholic Europe.

Catholics were far more active in investing the urban and rural landscape with new sacred buildings. Starting in Italy, the Iberian Peninsula and the Southern Netherlands in the late sixteenth century, followed in numerous places in Central Europe, Catholic clergy and territorial rulers responded to the rise of Protestantism with a massive drive to build new churches, cloisters, episcopal palaces, pilgrimage sites and other monumental structures that, in a splendid, exuberant style dubbed the 'Baroque', would represent and proclaim the power and the glory of the Catholic Church and faith.[196] While the completion of a sacred building during

193 James, *Template makers*, 33–35.
194 Shelby, 'Geometrical knowledge', 398.
195 Binding, *Bauwerk*, 101–107, Philip, 'Eyn huys in manieren van eynre kirchen', 73–74.
196 DaCosta Kaufmann, *Court, cloister and city*, 214–217, 343–345.

the Catholic Reformation typically took less time than the finishing of a Gothic cathedral, St Peter's Basilica in Rome being a notorious exception, the intense religious building activity in Catholic countries between 1600 and 1750 probably offered no fewer opportunities for informal learning than the great church-building projects between 1100 and 1500. Buildings in the Baroque age were, after all, a kind of *Gesamtkunstwerk*, which called for the cooperation of people with a wide variety of skills, from architects, masons, stonemasons, carpenters and joiners to plasterers, sculptors, painters, glaziers, organ builders and gold- and silversmiths. The Society of Jesus, arguably the most prolific founder of Baroque buildings, supplied not only architects from its ranks, but also carpenters, masons and stonemasons.[197] The 'design at St Peter's', Sarah McPhee has observed,

> was a cumulative process. Anyone who stepped into the role of architect inherited the visions, ambitions, and debates of generations of cardinals, popes and architects before him. Ideas survived in drawings, engravings and models, in the memories of men, and in the reflective memory and lore of the institution.[198]

Nevertheless, on the whole, religious institutions in China and Europe after the sixteenth century began to carry less weight in the development of informal learning, because opportunities for 'learning by doing' or 'learning by using' *outside* these institutions had become more plentiful than before. More alternative, durable settings for informal learning were available than had existed in earlier periods.

The most common of those durable, alternative settings were craft guilds, that is: permanent, government-recognized, local organizations of people from the same craft or trade whose purpose was to defend their shared interests.[199] Craft guilds – as distinct from organizations of merchants – have flourished for a very long time in both China and Europe. Such guilds probably arose earlier in China than in Europe. In the building industry, skills were transmitted by means of informal learning (aided by memorization in verse) in family guilds, which were regulated by the government under the Tang. In Song times, great cities in the Yangzi Delta, like Hangzhou, boasted a number of guilds in luxury crafts, such as

---

[197] Lombaerde, 'Introduction', 15–19, 25, Daelemans, 'Pieter Huyssens', 42, Fabri and Lombaerde, 'Architectural treatises', 187, DaCosta Kaufmann, *Court, cloister and city*, 214–217, 343–345.

[198] McPhee, *Bernini*, 36–37.

[199] Based on: De Munck, Lourens and Lucassen, 'Establishment and distribution', 32–34, 67.

jewellery making, hat making or fine cloth weaving.[200] From the twelfth century, craft guilds spread in Europe, too, first in larger and smaller cities in Italy, France, England, and the Southern Netherlands and later also in countless towns in many other parts of the Continent.[201]

There were also some remarkable differences between China and Europe, however. In Europe, thousands of craft guilds had been established by the 1550s, and their numbers expanded still further until the end of the eighteenth century. In the Northern Netherlands, for instance, nearly 60% of the c.1,400 craft guilds created between 1100 and the 1780s were founded after 1560. The total number of guilds (including craft guilds) known to have existed in China between 1550 and 1800, by contrast, is much lower, amounting to less than 150. Guilds in China did not proliferate more widely and rapidly until the nineteenth century, and the phenomenon did not spread equally across the entire empire. While most cities in the Yangzi Basin and the coastal area from Guangdong to Shandong saw a substantial increase in guilds, inland regions, frontier areas and a few places along the Yangzi or near the coast remained more or less 'guild-free zones'.[202]

Although cities were home to intense concentrations of crafts and trades, they never covered the entire gamut of technological activities. Both in China and Europe, many activities were wholly or partly conducted outside urban centres: farming, fishing and seafaring, obviously, but also mining, quarrying, iron production, textile-making, irrigation, dyke construction, canal building or inland transport. It was usually the case that non-urban occupations were organized in guilds to a much lesser degree than crafts or trades in cities. Rural guilds, or mixed urban-rural guilds, admittedly existed in large parts of Central Europe in the early modern period, especially in textile manufacturing,[203] but they were not very common elsewhere. Informal learning in non-urban occupations must have commonly taken place on site (on a field, on a farm, at sea, in a mine, at a quarry, near a furnace) rather than in a formal guild-supervised environment.

Both in China and in Europe, for a long time, craft guilds in cities combined economic and religious functions. Religious activities remained a

200 Guo, *Chinese architecture*, 90, Kuhn, *Age of Confucian rule*, 209–210.
201 Epstein, *Wage labor*, 50–91, Keene, 'Early history', 79.
202 Moll-Murata, 'Chinese guilds', 221–222, 243–244, 247, Lucassen, Lourens and De Munck, 'Distribution', 53–55.
203 Ehmer, 'Artisans', 67–68.

normal feature of guilds in China until the 1920s. Guilds were active in building, maintaining and embellishing temples from at least early Qing times onwards and they regularly convened for common religious worship. Craft guilds in Europe were also initially intimately connected to religious institutions. Members of craft guilds attended funerals of deceased brethren, prayed for the salvation of their souls, marched in processions, maintained chapels and altars, gave alms and contributed to funds to pay priests to say masses and perform other ritual services. Guilds often used churches as meeting places.[204] Many guilds may in fact 'have owed their existence to the original, common religious observances of members in the same craft in either a confraternity or the parish church', Steven Epstein has observed. 'Virtually every guild was also a confraternity', although the reverse was not always true.[205]

These very institutions also played a pivotal part in the supervision of training of new craftsmen. Craft guilds in Europe not only regulated the duration of apprenticeships, but also laid down the requirements for new entrants to a craft, checked their qualifications and determined who would be allowed to train them. Although craft guilds in Europe were never the sole agencies for the supervision of apprenticeships (and seldom had exclusive power to set the rules),[206] they must have exerted considerable influence on how informal learning in crafts developed in practice. Guilds in China made rules about apprenticeship, too, but 'primarily [as] a custom-sanctioned initiation ... before one could enter a trade'.[207] In contrast with Europe, Chinese guilds did not require examinations at the end of an apprenticeship term. The level of expertise to be reached at the completion of the apprenticeship thus was left undefined.

In a sense, the religious upheavals that shook Europe during the sixteenth and seventeenth centuries marked a break in the connection between crafts and religious practices. This was, understandably, more true for regions that joined the Protestant cause than for regions that remained in the fold of Catholic Church. In Protestant territories, such as the Dutch Republic and England, craft guilds were almost entirely shorn of their religious functions. Guild altars disappeared, masses for deceased

[204] Moll-Murata, 'Chinese guilds', 223, 229, 241, Epstein, *Wage labor*, 155–169, Richardson, 'Christianity and guilds', 145–152.
[205] Epstein, *Wage labor*, 156–157.
[206] De Munck, Kaplan and Soly, "Learning on the shop floor", 10–16.
[207] Golas, 'Early Ch'ing guilds', 566.

members stopped, processions and other religious celebrations by guilds no longer took place.[208]

In Catholic regions, guilds continued to take part in religious activities, but they did not do so everywhere on the same scale as before the sixteenth century. Crafts and confraternities were no longer as closely intertwined as they had been in the Late Middle Ages. Tridentine reformers encouraged the growth of confraternities, but they valued these organizations primarily as a means for enforcing moral discipline. A papal bull of 1604 declared confraternities exclusively 'ecclesiastical bodies', which meant that they needed formal approval from a bishop and came under regular supervision of clerical authorities. Their spiritual purpose was increasingly stressed at the cost of their convivial functions.[209] While the number of confraternities grew and their membership became more inclusive and diverse, guilds sometimes lost their vigour and relevance as organizing frameworks for religious activities. In the Spanish Netherlands after 1600, for example, guild members persisted in maintaining and embellishing altars and chapels, attending masses and partaking in processions, but their involvement in such collective guild-based activities was no longer as massive and regular as before. Guild members increasingly found solace for their spiritual needs in non-occupationally based religious fraternities.[210] A 'division of labour' between confraternities and guilds did not invariably occur, however. In Dijon, France, guilds and *confréries* remained firmly linked. All artisans who became masters were required to join a craft-based confraternity.[211]

In short, craft guilds across Europe continued to offer a crucial setting for informal learning until the end of the Old Regime (and in Central Europe, even until the mid-nineteenth century), but the connection between economic and religious activities after the Reform movements of the sixteenth and seventeenth centuries became somewhat weaker than before.

208 Thijs, 'Religion and social structure', 165–167, 173, Richardson, 'Christianity and guilds', 171–175.
209 Farr, *Hands of honor*, 242–246.
210 Thijs, 'Religion and social structure', 167–171.
211 Farr, *Hands of honor*, 243–244.

Conclusion

Despite their evident limitations, indicators of human capital formation used in present-day historical research nevertheless tell us a few things about developments in China and Europe between c.700 and 1800. The data available so far suggest that the *timing* and *direction* of trends in human capital formation in China and Europe in this period were more or less alike, but that differences may have existed in the *level* of these trends and in the way human capital was distributed in society. Information on literacy and, in particular, the number of book titles, seems to hint at this possibility. The extant evidence is further revealing in the sense that it does not confirm common assumptions that Protestant regions in Europe experienced a lead over Catholic ones, nor does it corroborate conventional suppositions about consistently growing differences between Northwest and Southern Europe. The trajectories of Spain or Italy do not easily fit into such conventional straitjackets.

Differences between China and Europe and within Europe itself become clearer, however, when we turn to the formal and informal ways in which in human capital formation actually took place. From the Song onwards, formal education in China tended increasingly towards a system composed of a multi-layered state school complex and a network of private schools. From this time in the schooling system, religious institutions were more and more relegated to a subordinate position. At best, temples were an adjunct of schools, not the other way around. In Europe, by contrast, religious institutions not only initially made a much larger contribution to the provision of formal education than state governments, but they also continued, or once again began, to be important suppliers of educational facilities even after secular institutions had also entered the field. From the sixteenth century onwards, religious institutions made a remarkable comeback in elementary, secondary and higher education. But differences within Europe increased, too, because the extent to and manner in which religious institutions were involved in education clearly varied by region. In territories in Europe which took the side of the Protestant Reformation, urban magistrates or territorial rulers assumed the functions that were formerly performed by monastic or cathedral schools. In regions in Europe where Catholicism continued to reign supreme, education in the early modern period came *more,* not *less,* under the control of religious institutions. In both Protestant and Catholic areas, the supply of educational provisions on the whole expanded rather than contracted.

A growing divergence in formal educational facilities between China and Europe and within Europe itself after c.1500 also manifested itself more specifically in the area of vocational and technical education. In China, schools with a vocational emphasis were few and far between until the middle of the nineteenth century. Prior to the nineteenth century, technical education in China largely took place via informal learning in the context of families, partly aided by the use of printed or written materials. In many places in Europe, by contrast, formal learning after 1500 branched out into vocational and technical schooling, albeit in different ways. While governments and/or religious institutions took the lead in Spain, Italy, France and Central Europe long before the nineteenth century, market forces and non-profit institutions were the decisive forces in England and the Dutch Republic.

For a long time, monasteries remained of paramount importance in informal learning in both China and Europe. While in China, these institutions gradually lost their influence from the twelfth century onwards, however, they continued to play a crucial role in many parts of Continental Europe until the end of the Old Regime. Moreover, the building sites of churches in Europe offered opportunities for informal learning to a greater number and variety of people than temple-building sites in China, especially in the period between c.1100 and 1500. Not until after the sixteenth century did religious institutions in Europe also begin to carry less weight in the provision of informal learning. By then, alternative opportunities for learning by doing and learning by using had become more plentiful than before. The most common of these alternative institutions were craft guilds; and after the Protestant and Catholic Reformations, craft guilds no longer displayed a religious identity to the same extent as before.

The spread and activities of religious institutions thus explain at least some of the variations in human capital formation relevant for technology between China and Europe and within Europe itself. Differences in human capital formation in China and Europe in the course of time cannot fully be understood without taking the impact of religious institutions into account. Religious institutions affected the manner and extent of human capital formation in different regions in different ways.

The roots of these disparities lay in the varying relations between religious institutions, governments, markets and demand 'from below'. In the interplay between these variables, religious institutions in Europe before 1500 could act more independently than in China. Even though the material resources of ecclesiastical institutions in Latin Christendom in this period may have been no larger than those of Buddhist or Daoist institutions in Tang, Song or Yuan China, the organizational capabilities were in all probability substantially greater. From the High Middle Ages onwards, religious institutions in Latin Christendom formed what Michael Mitterauer has called 'a highly organized religious community'. Between the late eleventh and early thirteenth centuries, the Western Church managed to achieve both a high degree of internal cohesion and a large measure of independence with regard to secular authorities. Organizationally, in this period, the Catholic Church was strengthened by the twin movements of what Harold Berman called the 'Papal Revolution' and the emergence of formally linked networks of monastic orders (discussed in Chapter 3), both of which embraced the whole world of Latin Christendom. The 'Papal Revolution' involved the establishment of the supremacy of the papacy over the entire Church and the emancipation of the clergy from control by secular authorities, as well as the creation of a nuclear papal bureaucracy and envoy system plus the constitution of the 'rule of law' by the formation of a new legal system, the 'canon law' of the Roman Catholic Church.[212] Thanks to these organizational structures and regulations, religious institutions in Europe were able to make, in a relatively autonomous and systematic way, a substantial impact in a wide variety of fields, ranging from morals and politics to economics and education. Clergy established and staffed many schools and institutes for higher learning. Although the independence of religious institutions vis-à-vis secular governments diminished from the sixteenth century, especially in Protestant countries, they could still make an active contribution to the expansion and maintenance of educational facilities in society at large. Sometimes they did so on their own accord, and sometimes their actions were a response to demand 'from below' or from secular authorities.

Reducing or abolishing the educational provisions offered by religious institutions was not necessarily detrimental for human capital formation, if their role was adequately taken over by other institutions. In Hesse-Kassel and the Netherlands in the sixteenth century and in Austria in the 1770s, for instance, governments reallocated part of the income generated by the seizure of ecclesiastical property for that very purpose. In these cases, the attack on religious institutions did not deal a severe blow to the infrastructure of formal learning as such. But the persistence of

[212] Mitterauer, *Warum Europa*, chapter 5, esp. 152–154, Berman, *Law and revolution*, 2, 85–119, 520–537, Mann, *Sources of social power*, 379–390, 503–507.

educational provisions maintained by religious institutions was far from harmful: through their role in elementary, secondary and higher education, religious organizations made a major contribution to human capital formation in Italy, Spain, France, the Southern Netherlands and parts of Germany in the seventeenth and eighteenth centuries. The education that they provided was often both good and cheap. It was not seldom assumed to offer the best value of its kind, especially at the levels of secondary and higher education. Religious institutions offered education that was considered to be of sound quality and easily accessible, because it was usually supplied for free. Formal instruction in religious institutions in Catholic Europe, as we have seen, comprised much more than rote learning of religious texts. On the contrary, in many places, these institutions were in the vanguard of the development of vocational and technical training. More than a few of them were leading centres in the teaching of mathematics. If, in the seventeenth and eighteenth centuries, a gap in economic performance emerged between areas in north-western Europe and other regions of the Continent, which in many cases happened to be Catholic, this 'Little' divergence can hardly be ascribed to a lack of educational facilities or to the prominent place of religious institutions in the provision of formal learning in Catholic territories.

CHAPTER THREE

RELIGION AND THE CIRCULATION OF TECHNICAL KNOWLEDGE IN CHINA AND EUROPE

Circulation of knowledge has become a key concept in the history of technology and the history of science. Circulation is said to take place once knowledge 'ceases to be the exclusive property of a single individual or group and becomes part of the taken-for-granted understanding of much wider groups of people'.[1] Kapil Raj has observed that circulation comes into play at two moments: 'firstly, in the collection and centralization of information or data from the outside world in some fixed place', and 'secondly, in the dissemination of the knowledge produced in this setting'.[2] Circulation of knowledge encompasses all the various aspects of the 'social history of knowledge' distinguished by Peter Burke (although his conception of knowledge primarily refers to 'academic', not technical, knowledge): professing, establishing, locating, classifying, controlling, selling and acquiring knowledge.[3] Circulation of knowledge has both social and spatial dimensions and can occur in many forms and ways and through many different carriers: people, goods, images, sounds or texts. In the course of circulation, knowledge and its carriers can change, too.[4] The concept of circulation does not presuppose a kind of hierarchy among the places between which flows of knowledge take place, although in historical reality, hierarchical relationships may nevertheless sometimes be imposed. Movements and communications of knowledge are not *a priori* viewed in terms such as 'centre', 'periphery' or 'transfer'. The 'mutable' nature of knowledge and its carriers, as well as the essential equality between localities, are crucial characteristics of the concept of circulation of knowledge.

Chapter 2 looked at localities of human capital formation between 700 and 1800, such as schools, monasteries and crafts, and the ways in which formation of human capital at these sites occurred. It discussed, more

---

[1] Secord, 'Knowledge in transit', 655.
[2] Raj, *Relocating knowledge*, 225.
[3] Burke, *Social history*, esp. 11–14.
[4] Hilaire-Pérez and Verna, 'Les circulations techniques', Belfanti, 'Guilds, patents', Raj, *Relocating knowledge*, 20–21, 226.

specifically, how religious institutions affected the formation of 'mental or physical equipment embodied in individual persons' that made them more capable in the use, improvement or innovation of technical artefacts and practices. In this chapter, the focus of the analysis shifts to the circulation between localities and the perspective on religion broadens to encompass both institutional and ritual and practical aspects. The key question of this chapter is whether, how, and to what extent religious institutions, rituals and practices in China and Europe between c.700 and 1800 shaped and maintained the circulation of technical knowledge as defined above.

Like the previous chapter, this chapter makes systematic comparisons between China and Europe and between different areas within Europe in order to understand the origins of the Great and Little Divergences. It consists of two sections. The first section analyses the parallels and differences in patterns of circulation of technical knowledge in China and Europe in the period up to about 1500, and the second section discusses the growing differences that developed thereafter. The early sixteenth century was a real watershed in many ways, and this watershed partly had to do with changes in religious context.

Both sections examine the role of religious institutions, rituals and practices in the circulation of technical knowledge. To pin down the impact of religious factors, each of these sections discusses movements of people, texts, images and artefacts, as well various ways in which information was collected, stored and disclosed; and then, more specifically, inquires as to where, when and how variations in religious context mattered. These are the issues to which I will now turn.

Religion and the Circulation of Technical Knowledge Before 1500

Patterns of Circulation

In the period up about 1100, fixed sites for the collection and storage of technical knowledge in China and Europe were still few and far between. Tang China and early medieval Europe had hardly any permanent spots where information on the methods, practices, mechanisms or ingredients used in technology was recorded and stored. As we have seen, monastic houses in both China and Europe were of prime importance for the preservation and spread of literacy, the production and reproduction of books and the continued development of informal learning. Libraries could be found in various places, mostly in monasteries, cathedrals or palaces. In

China, an imperial collection of books had existed since Han times.[5] There is no evidence, however, that these sites also functioned as centres where *technical* knowledge was described, copied or catalogued in such a way that it could be easily and widely shared and disseminated.

The prime movers in China in the recording and storing of information about technology-in-use were not religious establishments, but state agencies. Unlike Buddhist and Daoist institutions, which used woodblock printing only for the distribution of individual religious texts and canonical works,[6] government offices under the Song, Yuan and Ming dynasties published books about all kinds of technical topics as 'an aid for good government' and the promotion of 'the welfare of the general population'. Scholar-officials compiled dictionaries, anthologies, tracts and treatises full of textual and visual data on subjects such as agriculture, textile-making, building, carpentry, water management, paper-making, salt production and shipbuilding.

Books on agriculture (*nongshu*) and textile technology, in particular, 'were circulated to local magistrates so they could develop the rural economy in their district', something that was regarded as one of the most important obligations of the emperor and his officials.[7] In contrast with the brief *nongshu* composed by landowners from the twelfth century onwards, which mainly dealt with farm management, were only circulated in manuscript form and rarely travelled large distances, official *nongshu* were intended 'to produce comprehensive, mobile knowledge that could successfully be transferred through the medium of print, across the vast spaces of the empire, and translated into local action'.[8]

The *Genzhi tu*, compiled by a magistrate in the Yangzi Delta in the 1130s, for example, showed in a set of 45 paintings, accompanied by poems, the most important tasks of rice-farming and sericulture. It had a wide circulation in the Southern Song Empire.[9] Another treatise on agriculture and sericulture, called *Nongsang jiyao*, produced in 1279 under the aegis of the newly-founded Board of Agriculture of the Yuan dynasty, was distributed in print all over China. One of the highlights of this book was the new information it provided on a 'recent, improved method of cotton

5 Guy, *Emperors' Four Treasuries*, 12.

6 Chia and De Weerdt, 'Introduction', 11, 23, Blair, 'Afterword', 359, Huang, 'Early Buddhist illustrated prints', 134–137, Chia, 'Uses of print', 174–175, 201.

7 Bray, 'Introduction: the powers of *Tu*', 26, Bray, 'Science, technique, technology', 330.

8 Bray, 'Science, technique, technology', 331–334.

9 Franke, *Kêng tschi t'u*, 69–72, Bray, 'Agricultural illustrations', 521, 524–527.

cultivation'.[10] The *Nongshu* by Wang Zhen, completed in 1313, broke new ground by showing the details of the construction of tools, equipment and machinery used in farming and textile-making, and by integrating technical drawings and diagrams with texts. Among the illustrations of mechanical devices included in Zhen's *Nongshu* were drawings of a winnowing-fan, a silk-reel, water-powered bellows and irrigation equipment.[11] A separate treatise on looms was compiled in 1264.[12]

Agricultural treatises were not only important for the accumulation, standardization and dissemination of knowledge on agriculture as such, Kent Deng has argued, but they also 'functioned as a means to bridge technological gaps'. Without these books, he claimed, agrarian recovery after repeated natural or man-made disasters would have been much more difficult (even though not all *nongshu* have been preserved). 'In this sense, Chinese technological literature, of which the *nongshu* formed an important part, contributed greatly to the survival of the Chinese agriculture-based society', Deng concluded.[13]

Detailed information on building techniques was recorded in the *Yingzoa Fashi*, compiled by a superintendent of state buildings, Li Jie, in 1100, and published by the Song government in 1103. The *Yingzoa Fashi*, which was primarily intended as a manual for government officials to keep a check on expenditures of building activities, specified units of measurement, design standards, methods of construction and working procedures, gave a glossary of technical terms, provided data on labour costs, explained the manufacture of bricks, tiles, mortar, pigments and glazes and illustrated details of carpentry and joinery by means of drawings. It helped to standardize existing practices and to disseminate these standards throughout the empire.[14] The *Ziren yizhi* by Xue Jingshi, dated c.1260, and the *Lu Jang Bin*, compiled in the fifteenth century, were storehouses of information about machines and pieces of equipment. The *Ziren yizhi* contained, among other things, a description of the construction of looms, including illustrations and lists of parts and measurements. The *Lu Ban jing* provided explanations (plus lists of parts and measurements) for the making of hand- or foot-driven irrigation wheels. While the *Ziren yizhi* has not been preserved as a separate work, but only as part of an

---

10 Bray, 'Science, technique, technology', 335.
11 Bray, 'Introduction: the powers of Tu', 67, Bray, 'Agricultural illustrations', 535–543.
12 Kuhn, *Webstühle*, esp. XXVIII.
13 Deng, *Development versus stagnation*, xxii, 3–5, 176.
14 Guo, *Chinese architecture*, 90–96, Ledderose, *Ten thousand things*, 132–134.

encyclopaedia composed at a later date, the *Lu Ban jing* remained in circulation for a long time. Copies were still on sale in bookshops in China in the early twentieth century.[15]

River control, too, became a subject of publication from the Song onwards. Half a dozen works wholly or partly devoted to this topic appeared in the eleventh and twelfth centuries.[16] The oldest manual on paper-making dates from the late tenth century, the first book on salt production techniques from the 1330s.[17] Descriptions of ships and shipbuilding techniques were composed under the Song and the Late Ming.[18] Mining and smelting, however, remained outside the purview of writers on technical subjects. This was an area with which scholar-officials in China were hardly concerned.[19]

It was noted above that texts were sometimes combined with images. Although 'graphic images or layouts which encoded technical knowledge' (*tu*) must have existed at least since early imperial times, primarily as geographical maps, the range of fields that produced technical drawings and diagrams widened with the advent of print culture and became even broader from the Song onwards. After the tenth century,

> every field of technical knowledge began to generate its own *tu*, many of which rapidly entered general circulation, not least because printing greatly facilitated the production of encyclopedias, compendia and other collections that included illustrated materials from a range of fields of technical knowledge.[20]

In about 1150, the historian and philosopher Zhen Qiao suggested that a mixture of images and texts was essential for shaping understanding: 'the images (*tu*) are the warp threads and the written words (*shu*) are the weft'. Wang Zhen's *Nongshu* of 1313 realized this ideal to the full for the first time, and it is beyond question that his combined texts and images were widely diffused.[21]

---

[15] Ruitenbeek, *Carpentry*, 1, 31–32, 270–276, Deng, *Development versus stagnation*, 38–42, Needham, *Science and Civilisation*, vol. IV part 2, e.g. 153–154, 204, 371–373.

[16] Needham, *Science and civilisation*, vol. IV part 3, 324–325.

[17] McDermott, *Social history of the Chinese book*, 13, Yoshida, *Salt production techniques*, 2–3.

[18] Scheuring, *Drachenfluß-Werft*, 2, 5–6.

[19] Vogel, 'Mining industry', 177–178.

[20] Bray, 'Introduction: The powers of *Tu*', 4, 13, 22, 26.

[21] Bray, 'Introduction: The powers of *Tu*', 38–40, Bray, 'Agricultural illustrations', 537–543.

Images did not only circulate in print, however. The *Yingzoa Fashi* documents that by the early twelfth century, drawings in different forms could serve as aids in various stages of a building process, although it is difficult to establish how frequently they were used in practice. These architectural drawings were made with brush and ink on silk or paper, or were applied by paint on walls, stones or brick. Working drawings displayed the site of a building (or of a complex of buildings), its grid at ground level, its elevation (by means of a cross-section or a longitudinal section), its front view or its aspect from above. Presentation drawings showed how a building, or a set of buildings, looked in its finished state. Architectural models were in circulation as well. Like working drawings, paper working models could function as an aid in the design and construction stage. Presentation models, made of wood or paper, were used in communication between builders and clients.[22]

In Europe, technical information was not recorded in print until the end of the fifteenth century and the prime movers in collecting this knowledge were not state agencies, but individual craftsmen, engineers, physicians and humanists. The extent of centralization of information was certainly much lower than in contemporary China. Yet, the mere fact that from the twelfth century, a single style of architecture, 'Gothic', could spread to every corner of Latin Christendom suggests that technical knowledge nevertheless must have travelled widely and that the degree of standardization, in spite of the decentralized circumstances, was in practice much higher than the absence of printed technical manuals, tracts or encyclopaedias might lead one to suppose. Aside from the mobility of labour, which I discuss below, one of the explanations is the increased circulation of knowledge via templates, drawings, models and manuscripts.

Templates are known to have been used in cathedral building from at least the early thirteenth century onwards. Templates (*formas, patrons, molles, mallen* or *brederen*) were forms carved from wood or made from paper, linen or metal, which allowed stonemasons to cut stones exactly according to the profile needed for a specific part of a building. Templates were designed by the master mason in charge of a building project, using his particular geometrical rules of thumb, and then distributed to workmen on the spot, or sent to stonemasons in a distant quarry.[23] Working

---

22 Guo, *Chinese architecture*, 115–124.

23 James, *Template makers*, 33–36, Shelby, 'Geometrical knowledge', Binding, *Baubetrieb*, 229–234, Philipp, 'Eyn huys in manieren van eynre kirchen', 77–78.

drawings emerged in stages from the Early Middle Ages onwards. Site plans in script existed as early as the ninth century. Drawings of windows, front views, cross sections and grids at ground level first appeared in stone and on parchment in about 1250 and on paper in the early fifteenth century. Other kinds of designs and plans (*visieren, visirungen, patroenen*) on paper or on parchment made their appearance in the late fifteenth century.[24] Except for the plans carved in stone, all these drawings could easily travel from one place to another. Another portable form of knowledge was that contained in the three-dimensional models of buildings used for presentation purposes, or as an aid in the design and construction phase. The use of models commenced in Italy in the fourteenth and fifteenth centuries and spread north of the Alps after 1500.[25]

Manuscripts with technical information in verbal and pictorial form, by contrast, were quite rare in Europe before the fifteenth century. Master mason Villard de Honnecourt from Picardy did something very unusual in about 1220 when he collected a large number of drawings and explanations of building techniques and mechanical devices (such as a sawmill) in a sketchbook. A tradition of writing on technical subjects by craftsmen did not yet exist.[26]

Writings on such topics did not become more common until after 1400. There was 'a veritable explosion of technical treatises' by artisans between 1405 and 1420. This sudden expansion in technical writing was not primarily related to the production of ecclesiastical buildings, but rather had to do with changes in warfare, the growth of commerce and the rise of patronage by secular rulers.[27] Master gun-makers in Germany in the early fifteenth century began to record information on their art in notebooks in the vernacular, illustrated by drawings, intended for communication with fellow craftsmen.[28] In the late fifteenth and early sixteenth centuries, booklets with information on existing building practices were composed by master masons like Matthias Roriczer, Hanns Schuttermayer and Lorenz Lechler.[29] Engineers, military physicians and humanists in northern and central Italy and in southern Germany began to write treatises in

---

[24] Binding, *Baubetrieb*, 171–188, 191–206, Philipp, 'Eyn huys in manieren van eynre kirchen', 88–89.

[25] Binding, *Baubetrieb*, 188–190.

[26] Shelby, 'Geometrical knowledge', 408–411.

[27] Smith, *Body of the artisan*, 31, Hall, "Der Meister sol auch kennen schreiben und lesen", Leng, 'Social character', 88–89, 106.

[28] Leng, 'Social character', 87–90.

[29] Shelby, 'Geometrical knowledge', 408–420.

Latin or in the vernacular, addressed to princely or urban patrons, on all sorts of topics in the mechanical arts. These treatises, which founded the genre of 'machine books', partly gave a description of technology-in-use, partly projected images of new machines or contraptions that in future might usefully be brought into practice in warfare, water-lifting, transport, building or industrial production. Examples of imagined devices abound in the notebooks of the engineer/architect/painter Leonardo da Vinci, but they also appear in writings by other authors composed after 1400, such as those by Konrad Kyeser or Mariano Taccola. 'Machine books' presented mechanical devices, such as mills, winches, cranes, pumps, ships and engines of war, by means of visual representations and verbal explications. Except for the booklets by Roberto Valturio and Matthias Roriczer published in the 1470s and 1480s, such writings on technical subjects circulated only in script.[30] The distribution of these products of the 'last scribal age' could nevertheless be quite broad. Konrad Kyeser's manuscript *Bellifortis* (1405), for example, has been preserved in at least 45 copies, almost all of them dating back to the fifteenth century.[31]

While technical information in China and Europe thus sometimes travelled by means of objects or texts and images in script or print, most of the circulation of knowledge must have occurred via mobility of people. Mobility of people in China was, in a sense, institutionalized by the state via the system of civil service examinations introduced under the Tang. The examination became more stratified in the course of time. While examinations in the Tang period were only held in the imperial capital, they were expanded to two levels under the Song (one in the prefectures and one in the capital) and to three under the Ming: one at the county, department and prefectural level, one at the provincial levels and one at the level of the metropolis. Examinations for the first levels took place once a year in local centres, while those for the higher levels were conducted every three years.[32] Thus, at regular intervals, large numbers of people were on the move to central places where examinees and examiners would meet. The size of these flows of travellers grew as the number of candidates increased. While in Tang times no more than a few hundred people sat an exam each year, the numbers under the Song ran into the hundreds of thousands. The annual number of participants in the lower,

---

[30] Long, *Openness*, 102–142, Leng, 'Social character', 86.
[31] Lohrmann, 'Das Maschinenbuch', 74, Long, *Openness*, 102, 105–109.
[32] Elman, *Cultural history*, 133.

prefectural exams rose from c.100,000 in the middle of the twelfth century to 400,000 or more a century later. By the middle of the eleventh century, examinations for the highest degree drew between 5,000 and 10,000 candidates to the imperial capital every three years. For each candidate, this meant 'several months' of travelling on roads and barges and of staying at inns or guesthouses in the capital'.[33] Whether candidates for examination seized the opportunity to gather or pass on technical knowledge en route is impossible to tell. As would-be scholar-officials, they may have reasoned that it might be useful to be informed about technical practices, although they were not expected to do manual labour themselves.[34]

A more direct way for technical knowledge to circulate was the movement of skilled labour itself. The extent of this movement is difficult to gauge, but it is certain that it took place and it is even very likely that it accelerated when building activities increased under the Song. A striking contrast between China and Europe before the mid-sixteenth century was that for a long time, movements of skilled labour in China were subject to a high degree of central regulation. Under the Tang and Song dynasties, tens of thousands of artisans were brought under control in the Palace Domestic Service and the Directorate for Imperial Manufactories. Artisan families were called upon to serve the state, a practice that was continued under the Yuan.[35] Such regulations did not necessarily obstruct the craftsmen's movements, but they implied nevertheless that the circulation of skills was tightly controlled by the government.

Under the early Ming, these regulations were reintroduced and expanded into a full-blown corvée system. Various categories of producers were henceforth obliged to render regular labour service to the state. The corvée system was based on an elaborate system of registration, combined with strict rules that restricted changes of occupation and set narrow bounds to physical mobility. Farmers, soldiers, merchants or artisans were not allowed to switch to another profession and people could not travel long distances without official permission. According to an occupational census taken in the early Ming, the empire then had nearly 300,000 artisans. These artisans were divided into two groups: 'resident artisans', who lived in or near the capital and had to report for duty in the palace workshops for ten days each month, and 'shift artisans', who

33 Kuhn, *Age of Confucian rule*, 121–123.
34 Cf. Elman, *Cultural history*, 482–483, Will, 'Apprentissage', 10–11.
35 Guo, *Chinese architecture*, 90, Moll-Murata, *State and crafts*, 32–39.

lived and worked everywhere in the country and had to render service for a number of months in the capital once every few years (fixed at four years in 1445). Only on those occasions were they allowed to travel further than the normal limit of 100 li (58 kilometres) from their place of residence.[36] Shift artisans made up 80% of all registered artisans. Of this category of artisans, half were active in the building trade, including, for instance, ten thousands of carpenters and thousands of sawyers, masons and bricklayers.[37]

Klaas Ruitenbeek has suggested that, for example,

> for a carpenter living in a remote corner of the empire it must have been quite an experience to travel to the capital and meet with fellow craftsmen from other provinces. To him, it may have meant something comparable to the *tour de France des compagnons* or the *Wanderjahre* of the artisans in Europe.[38]

If this were true, it would mean that knowledge and skills in many crafts and trades could circulate constantly and widely across the entire empire. It is doubtful, however, whether the labour service system really fulfilled this function of state-sponsored circulation pattern equivalent to the practice of tramping in early modern Europe. Ruitenbeek himself admits that 'it remains unclear... to what extent, and to what distance from the capital, the artisans of the empire were affected by the corvée system'.[39]

However, both during times when a system of labour service and/or registration was in operation and during the periods in between, movements of skilled labour must have also continued in other sorts of ways. From the late Tang until the collapse of the Northern Song in about 1130, the border region of Hopei and Honan saw an astonishing expansion of iron mining, iron-making, metalworking and, especially in the eleventh century, coal mining. The whole area came to be densely settled with mines, forges and furnaces, which produced a yearly amount of iron, coal and metal tools and implements that was not matched anywhere in the world until the beginning of the Industrial Revolution.[40] Although Robert Hartwell fails to mention it explicitly in his enumeration of economic and political

---

36 Friese, *Das Dienstleistungs-System*, 117–130, Ruitenbeek, *Carpentry*, 16, Brook, *Confusions of pleasure*, 19.

37 Ruitenbeek, *Carpentry*, 16–17.

38 Ruitenbeek, *Carpentry*, 18.

39 Ruitenbeek, *Carpentry*, 18.

40 Hartwell, 'Cycle of economic change', 106–145 and Donald Wagner in Needham, *Science and civilisation*, vol. V part 11, 278–325.

conditions that favoured the expansion of these industries in northeast China,[41] it is evident that a sufficient supply of skilled and mobile labour must have been a crucial element in this. How could innovations such as the use of coke-fuelled blast furnaces and direct de-carbonization in steel-making during the eleventh century have diffused throughout the region, if skilled workers had not been able to move from one site to another, given the fact that a body of technical writing on these subjects was not yet available? And the probability that this really happened becomes the greater if one keeps in mind that labour in the ferrous metal enterprises before the middle of the thirteenth century still enjoyed free status.[42]

Movements of skilled labour must have been an essential factor in the circulation of knowledge on machine making, too. The publications on technical subjects from the Song, Yuan and Ming dynasties that I discussed above reveal that China, by that time, already had a great variety of machines. These machines were usually wooden or bamboo devices with a number of revolving or spinning components, which could perform productive tasks in mining, agriculture and industry. Among the machines employed in China in this period were winches for mine-shafts, animal-powered grinding mills, human-driven winnowing-fans, hand-driven chain pumps, man-powered capstans for pulling vessels, scoop-wheels driven by humans or animals for lifting water into fields, water-lifting devices driven by windmills, water-powered bellows and water-driven multiple frames for spinning hemp.[43] Making such machines was not necessarily a specialized activity. Machine-building may have been a household activity, or sometimes a sideline of carpentry or the blacksmith's trade. Looms in the silk industry, for example, may have either been constructed or repaired in the households of weavers or 'by loosely attached but specialized carpentries'.[44] Yet, it is unlikely that local carpenters, blacksmiths or other craftsmen always possessed the entire range of skills to construct whatever mechanical device was required. A more plausible supposition is that machines were often built by craftsmen travelling from one place,

41 Hartwell, 'Cycle of economic change', 154.

42 Hartwell, 'Cycle of economic change', 150.

43 Needham, *Science and civilisation*, vol. IV part 2, *passim*, esp. table 56, Zang, 'Ancient Chinese windmills', 203–204, Xiao-Wu, 'Interpretation', *passim*, Leonard, *Controlling from afar*, 16–17.

44 Schäfer, 'Silken strands', 66–67, cf. also the treatise on making looms in the mid-thirteenth century edited and discussed by Kuhn, *Die Webstühle*, 7–8.

or region, to another. At any rate, special terms for artisan-engineers, such as *jiang* and *gong*, existed at an early date.[45]

Like in China, some groups of skilled workmen in Europe circulated beyond the confines of individual villages and towns and their immediate neighbourhoods. Notably, from an early date, machine-making became to some degree a specialized activity. Experts in the construction of machines came to be described with special terms such as *artifex*, 'engineer', *ingeniator, ingegnere, ingeniero, Werkmeister.*[46] Almost from their very first appearance in eleventh-century sources, these 'engineers' or *ingeniatores* were a highly mobile group. They made long-distance trips all over Europe to hire out their services to territorial princes, church authorities, urban governments and, finally, states, for the construction of building equipment, siege engines, industrial mills, clocks, water-lifting devices and sundry other ingenious mechanisms.[47] Master gun-makers in the late fourteenth and fifteenth centuries used to move from town to town, from or court to court, too.[48] Master masons or clock makers often travelled to other places to inspect a particular building or mechanical device *in situ*, or to oversee some complex construction project. It was not uncommon for a building owner or the awarding authority of a project to request that the craftsman contracted to execute the job model a new church, tower, city hall or timepiece specifically on the *forma* or *scema* of a similar object in another town.[49] After about 1350, mobility of craftsmen became even more common as the practice of tramping (*Wanderschaft*) began to spread throughout Central Europe. Tramping meant that journeymen – skilled workmen who had completed a guild-regulated apprenticeship – were required by custom, if not by local statute, to spend a year or more travelling between cities outside their home region. By the end of the sixteenth century, compulsory tramping became the rule in the Holy Roman Empire wherever crafts were organized into guilds.[50]

---

[45] Needham, *Science and civilisation*, vol. IV part 2, 9–10.

[46] See for the various names e.g. Dohrn-Van Rossum, 'Migration-Innovation-Städtenetze'.

[47] Dohrn-van Rossum, 'Migration-Innovation-Städtenetze', *passim*.

[48] Leng, 'Social character', 88–89.

[49] Binding, *Baubetrieb*, 179–185, Philipp, 'Eyn huys in manieren van eynre kirchen', 70–72, Icher, *Oeuvriers*, 63.

[50] Reith, 'Circulation of skilled labour', 116–119.

Religious Travelling

Religious institutions, rituals and practices likewise generated all sorts of medium- and long-distance movements of people, which may have facilitated the circulation of technical knowledge and skills across regions. People went on pilgrimage, for example, or they visited religious feasts, made missionary journeys, travelled to meet with fellow clerics in distant regions, carried out diplomatic missions on behalf of clerical authorities or joined in military expeditions sanctioned or inspired by religious beliefs. Travelling typically offers opportunities to get to know people, practices and objects in other places which embody knowledge and skills that are different from those at home, or that are even entirely new, and this applies to religious travelling as well. Whether religious travelling actually contributed to the transformation of knowledge produced in a particular environment from being 'the exclusive property of a single individual or group into the taken-for-granted understanding of much wider groups of people',[51] as the concept of circulation of knowledge implies, is difficult to tell, however. It is easier to detect or suspect a pattern of travelling than to discover the contents of information carried by travellers, although, as we shall see below, travel accounts can sometimes help us to get a glimpse of what they absorbed and conveyed.

If we assume that travelling can aid the circulation of knowledge, however, then it is plausible to suppose that travelling generated by religious institutions, rituals and practices did not play an insignificant part in that process. Religious travellers may, in several ways, have added something to the contribution of other categories of travellers such as merchants, warriors, or seafarers, or, in China, candidates for civil examinations. On the one hand, some groups of religious travellers, such as monks, must have been especially well-equipped to collect and register technical information, because they had acquired a singular set of abilities to do so. Notably, they were more literate than most other sections of the population, and they represented, to a high degree, human capital in motion. On the other hand, there were also types of religious travelling, such as making a pilgrimage or participating in a religious celebration, that involved sections of the population that would not normally have travelled great distances from home. Religious travelling allowed them, literally, to see more of the world than would have been the case in normal circumstances.

[51] Secord, 'Knowledge in transit', 655.

After c.1000, regular, long-distance travelling by monks probably became more common in Europe than in China. This had to do with the structure of monasticism. Starting with the wave of monastic reform spearheaded by the abbey of Cluny in eastern France, monasticism in Europe assumed, more than in China, the shape of formally linked networks. The network created by Cluny was extremely hierarchical. From the tenth century, Cluny became the 'mother-house' of a vast assemblage of 'dependencies' spread across France, Italy and Spain. Not only did all of these dependent monasteries copy the interpretation of the Benedictine practices introduced by Cluny, but they also recognized the abbot of Cluny as their supervisor, they accepted the right of the mother-house to appoint its superior and they even paid it an annual tax:

> All Cluniac monks, wherever they lived, were in theory members of the community of Cluny itself. When they visited the mother-house, they were entitled to take their place in chapter. Postulants who had been admitted to any house of the congregation ... were sent to make their profession to the abbot of Cluny or alternatively awaited the visitation of their own monastery.[52]

Monastic orders that were founded after the rise of the Cluniac movement assumed a more federal structure. The Cistercians, originating in Cîteaux in eastern France in about 1100, were the pioneers of a system of filiation and mutual supervision. The Cistercian order consisted of a number of families, comprising a founding abbey and a number of daughter-houses, whose abbots carried out a supervisory visitation of each other's institutions at least once a year, and participated in the election of a new abbot in an affiliated monastery where the headship had become vacant. All abbots, or their representatives, attended the general chapter of the order held in Cîteaux each year. Cistercian monasteries spread all over Europe. In the mid-twelfth century, the number of Cistercian monasteries already amounted to 343. By 1500, it had grown to no less than 1312: 738 for men and 654 for women.[53] Robert Friedel has suggested that the Cistercian order in particular, precisely because of its system of institutionalized mutual visitation and consultation, was of crucial importance for the regular circulation of knowledge in Europe. 'The order represented a network of expertise, exchanging personnel and experience throughout Europe, and in this way knowledge of productive ways of doing things

---

[52] Lawrence, *Monasticism*, 86–98, Wollash, 'Monasticism', 174–180, 183–185.
[53] Lawrence, *Monasticism*, 182–183, 186–192.

spread more effectively than they had in the early Middle Ages', he argued. 'Exchange of technical expertise among churchmen' thus helped technological advance in general.[54]

We will probably never have detailed knowledge of whether monks from different Cluniac or Cistercian abbeys exchanged much technical information during their frequently-scheduled formal meetings. Still, given the fact that monasteries were definitely important centres of agrarian and industrial production, the scenario is not implausible. And in the case of complex hydraulic technology (conduits, cisterns, pipes and fountains), there are quite a few examples of the diffusion of innovations between monastic houses, both in the Cluniac and in the Cistercian networks, partly thanks to the filiation relationship between houses and institutionalized consultation and visitation arrangements within particular monastic networks. The Cistercians even 'seem to have had something of a policy of sending out architects to help instruct the members of new foundations'.[55] In 1133, Saint Bernard of Clairvaux sent a senior monk from his monastery to a new community in Fountains to instruct its members not only in Cistercian customs, but also in architecture and the craft of plumbing.[56] Religious houses, in their turn, 'functioned as vital communication channels of the diffusion of awareness' of innovations in hydraulic technology to medieval townsmen'.[57] Because of their formal network structures, monastic houses in Europe between c.1000 and 1500 may have made a larger contribution to the circulation of technical knowledge than monasteries in Tang, Song or Yuan China. Although Buddhist or Daoist monasteries in China arguably differed in status or reputation and, especially from the Song onwards, also showed variations in practices by 'schools', 'lineages' or 'traditions',[58] they did not develop similar institutionalized hierarchical or federal structures to those of Cluniac or Cistercian networks in Europe.

Non-clergy could make a journey for religious reasons as well, however. Pilgrimage sites or religious festivities could motivate people from all sorts of backgrounds who did *not* form part of the monastic population to make longer trips or to travel to different places than they were normally wont

---

[54] Friedel, *Culture of improvement*, 21; a similar argument in Böhm, 'Technische Bildung', 90.

[55] Magnusson, *Water technology*, 13.

[56] Magnusson, *Water technology*, 11–13, 17.

[57] Magnusson, *Water technology*, 18–19.

[58] Ter Haar, 'Buddhist-inspired option', 95–96, Kohn, *Monastic life*, 66–67, Foulk, 'Myth', 163–167.

to do. Often they did so on a voluntary basis, sometimes as a compulsory form of penance. In Song and Yuan China, temple fairs on religious holidays frequently drew large crowds to cities. In the countryside, on feast days of local deities, worshiping societies gathered people from several villages to celebrate together in a common temple of their district.[59] Similar scenes took place during pilgrimages and other special devotional occasions in towns and villages throughout late medieval Europe. Apart from the 'national' shrine in Canterbury, England boasted dozens of regional or local sacred places which persisted to attract vast numbers of devotees up to the very eve of the Reformation. In April 1538, Henry VIII's vice-regent in spirituals, Thomas Cromwell, received reports that 'fyve or syxe hundrethe' pilgrims visited the shrine of Darvelgadarn on a single day. Communal processions and confraternity parades on saints' days continued to be hugely popular events in Italy. Organized pilgrimages by confraternities, as distinct from the customary pilgrimages by individuals, appear to have increased during the sixteenth century.[60] In contrast with China, war expeditions for religious reasons were for a while quite common in Europe (and in the Islamic world), too. Between the eleventh and fifteenth centuries, tens of thousands of people travelled in crusader armies between Western Europe, the Levant and other regions at the borders of Latin Christendom.

As far as medium- and long-distance pilgrimage to or from China before the sixteenth century is concerned, quantitative data are, alas, entirely lacking. The few extant travel accounts nevertheless show beyond doubt that such travelling by Buddhists *did* occur. In the late seventh century, several Buddhist monks from India made a pilgrimage to Wutai Shan (the mountain range with the Five Terraces) in Shanxi, China.[61] En route to the court of the Tang emperor in 839, the Japanese Buddhist monk Ennin visited this site, too, and further travelled 'to the estate of the Ting-chüeh-ssu [monastery]'.[62] From about 200 AD onwards, Chinese monk-pilgrims travelled by land to India, 'to seek instruction in Buddhist teachings, find scriptures, and visit famous sites sanctified by the life of the Buddha'. The most illustrious of these, Xuanzang, took sixteen years to complete his trip

[59] Von Glahn, 'Towns and temples', 189–190, Ter Haar, 'Buddhist-inspired option', 97.

[60] Duffy, *Stripping of the altars*, 191–195, 385, Webb, *Pilgrimage in medieval England*, 141–179, Black, *Italian confraternities*, 108, 117.

[61] Liu, *Silk and religion*, 26–27, Schneider, 'Un moine Indien', 31.

[62] Reischauer, *Ennin's diary*, 211–215, 267.

(629–645) and left an extensive account of his journey.[63] Given the subsequent shift in the centre of gravity of Buddhism from India and China to Nepal, Tibet, Japan and Southeast Asia, the pilgrim traffic between Japan, China and India probably became less frequent from about the thirteenth century onwards, but it may have intensified in other areas. And due to imperial patronage of 'Lamaist' Buddhism from the sixteenth century, China *did* attract pilgrims from Mongolia and Tibet.[64]

Although exact figures on pilgrimages by Christians are not available either, data about their spread and frequency are more abundant than in the case of China. It is known that Christian pilgrims began to travel to the Holy Land and other places in the Middle East (such as Mount Sinaï and various sites in northern Mesopotamia) in the second century AD.[65] This flow of religious travelling continued after the Muslim conquest of Palestine and Syria in the mid-seventh century. Michael McCormick has found that the numbers of pilgrims travelling from Italy, the Carolingian Empire, England, Ireland and Spain to the East rose in the later eighth and ninth centuries; the majority of these pilgrims were probably clerics. The most remarkable change in the pattern of travelling occurred with regard to the routes: 'Sometime between c.753/76 and 800/25, Western travellers heading for destinations beyond Constantinople abandoned the ancient Byzantine route and began to travel east through the Islamic world.'[66] Pilgrims from the west now preferred the route through Egypt over the route through the Byzantine empire.

The Crusades did not spell a major break in Holy Land pilgrimage from Western Europe either. After Jerusalem and the other holy places had returned to Muslim rule in the late twelfth century, Christian pilgrims continued to flock to the East – and they were no longer mainly clerics.[67] The flow of Christian pilgrims travelling to holy sites in Palestine did not begin to slacken until the sixteenth century. From then onwards, an increasing number of Christians who visited the Holy Land did so merely as 'one stage in a larger itinerary across the Levant'. These visitors were merchants, rather than pilgrims.[68] Long-distance pilgrimage in Europe

---

[63] Beal (ed.), *Buddhist records*, vol. I, ix–xx, Naquin and Chün-fang Yü, 'Introduction', 14–15.
[64] Naquin and Chün-fang Yü, 'Introduction', 20.
[65] Hunt, *Holy Land pilgrimage*, esp. 3–4 and chapters 3 and 4.
[66] McCormick, *Origins of the European economy*, 173.
[67] Peters, *Jerusalem*, 369, 423, Webb, *Medieval European pilgrimage*, 28.
[68] Peters, *Jerusalem*, 510, Peters, *Jerusalem and Mecca*, 44.

itself, especially to Rome and Santiago de Compostela, meanwhile continued unabated.

Studies on material life in China have made a reasonable case for the thesis that such long-distance religious travelling contributed to the spread of awareness of novel technologies, even if it did not involve the spread of actual know-how about their production. Xinru Liu has documented how pilgrims from Tang China to India carried large quantities of silk with them in order to buy Buddhist relics, and how visiting monks from India received silks or silk ritual robes as a donation or a reward from the Chinese emperor. This was one of the ways in which connections between silk and religious institutions, rituals and practices helped to transform silk between the seventh and eleventh centuries from an exclusive 'status symbol' of emperors into a 'common commodity' in 'many parts of Eurasia'.[69] Religious travelling could also help to spread knowledge about innovations that did not carry a symbolic meaning themselves. According to John Kieschnick, the introduction of the chair and the technique of making sugar to China, for example, can be understood as a by-product of the spread of the Buddhist monastic lifestyle and increased travelling by monks between China and India.[70]

Pilgrims and other religious travellers in Latin Christendom have left more written traces of their journeys than religious travellers in China.[71] While at the time, some of these travel accounts did not circulate outside the circle of the author's family, friends or fellow monks, many of them *did* reach a wider public, either in the form of a separate publication or in an indirect way, by being included or digested in a publication by somebody else. The *Evagatorium* by the Dominican monk from Ulm, Felix Fabri, for example, which described his two journeys to the Holy Land in the 1480s, appeared in an abridged German version (the *Pilgerbuch*) in the 1550s.[72] The 'history of the Mongols' by the Franciscan friar John of Plano Carpini, based on his observations as a papal envoy to the court of the Great Khan in 1245–1247, was soon after its completion incorporated into the popular *Speculum historiale* by Vincent of Beauvais.[73] William of Rubruck's account of his trip to Mongolia in the 1250s was not published, but his observations were partly digested into the *Opus maius* by friar

---

[69] Liu, *Silk and religion*, 2, 69–72, 195.
[70] Kieschnick, *Impact of Buddhism*, 220–222, 256–258, 275–280.
[71] On China: Eggert, *Vom Sinn des Reisens*, 66–68.
[72] Carls (ed.), *Felix Fabri*, 59–61.
[73] Dawson (ed.), *Mongol mission*, xv–xviii, 2.

Roger Bacon, who knew his story at first hand.[74] Moreover, pilgrims like Felix Fabri, who wrote accounts of their own journeys to the Holy Land, were often perfectly acquainted with narratives that had appeared in the past, or with pilgrim books produced by contemporaries.[75]

Given the purpose of their journeys, the accounts of these travellers were normally more informative on the religious experiences of their authors or on the relevance of their activities and observations for religious organizations than on the state of affairs in the surrounding world as such. Moreover, in accordance with the tacit or formal rules of the genre, the records of their travels were often cast in a traditional format, including all the *topoi* and stylistic devices that went with this, rather than being composed in a detached, open-minded fashion.[76] Nevertheless, these sources sometimes contained some information about technology, too, because religious travellers, in the course of their journeys, whether intentionally or not, happened to encounter people or to make observations in foreign places that provided information on ways in which other cultures or societies perceived their environment and went about in solving technical problems – and they did not fail to record these encounters or observations in their letters, accounts, reports or histories.[77]

Information on such topics could be presented in various forms and shapes. It could be let slip in casual manner, but it could also be furnished by means of lengthy descriptions and systematic expositions. In the account of his visit to the Ting Jue Si, for example, the monk Ennin incidentally remarked that 'he saw the water-[powered] mill'.[78] In the narrative of his journey to the court of the Great Khan in Mongolia, William of Rubruck observed that 'the ordinary money of Cathay is a piece of paper made out of cotton, a handbreadth in width and length, and on [which] they stamp lines like Mangu's seal'.[79] More extensive descriptions can, for instance, be found in Chinese monk-pilgrim Xuanzang's account of his travels in India in the seventh century:

---

[74] Dawson (ed.), *Mongol mission*, 88.

[75] Feilke, *Felix Fabris Evagatorium*, 7–9, Carls (ed.), *Felix Fabri*, 12–19.

[76] See for example the literary analysis of Fabri's pilgrim's book by Feilke, *Felix Fabris Evagatorium*.

[77] Cf. Magnusson, *Water technology*, 9, on the relationship travelling to Rome (including pilgrimage) and observation of hydraulic technology in Italian cities.

[78] Reischauer, *Ennin's diary*, 267.

[79] Dawson (ed.), *Mongol mission*, 171.

> The earth being soft and muddy, the walls of the towns are mostly built of brick or tiles. The towers on the walls are constructed of wood or bamboo; the houses have balconies and belvederes, which are made of wood, with a coating of lime or mortar, and covered with tiles. The different buildings have the same form as in China: rushes, or dry branches, or tiles, or boards are used for covering them. The walls are covered with lime and mud, mixed with cow's dung for purity.
>
> The garments [of the people] are made of *Kiau-she-ye* (kauśêya) and of cotton. *Kiau-she-ye* is the product of the wild silkworm. They have garments also of *Ts'o-mo* (kshauma), which is a sort of hemp; garments are also made of *Kien-po-lo* (kambala) which is woven from fine goat-hair; garments are also made from *Ho-la-li* (karâla). This stuff is made from the fine hair of a wild animal: it is seldom this can be woven, and therefore the stuff is very valuable, and it is regarded as fine clothing.
>
> With respect to the different kinds of wine and liquors, there are various sorts. The juice of the grape and sugar-cane, these are used by the Kshattriyas as drink; the Vaiśyas use strong fermented drinks; the Śramaṇs and Brâhmaṇs drink a sort of syrup made from the grape or sugarcane, but not of the nature of fermented wine.[80]

And Arnold von Harff from Cologne, travelling in the 1490s through Egypt on pilgrimage to the Holy Land, gave quite a systematic exposition of what he saw along the River Nile, not far from Rosetta:

> In these towns there grow many reeds or canes from which sugar is made. They have in each town a large building in which they prepare the sugar in this manner. They cut the cane, which is very sweet, down to the ground, and cut it into little pieces about the length of a finger or less. These they throw it into a great wide stone vessel, in which is a large mill stone for grinding, which oxen turn round on the lower mill stone, so that the stone grinds the sugar quite small. They then take the ground canes and empty them into a great long kettle holding nine or ten pails full, and light a fire underneath, so that it is seethed as they seethe saltpetre in our country, and skim off the top which they pour into tubs or barrels. This is called sugar honey, with which in this country all kinds of food are cooked, since they have no butter, which melts on account of the great heat. What remains at the bottom of the kettle they empty into pointed tubes where they let it cool. This settles and becomes the sugar loaf which is brought to our country in great quantities.[81]

Of all of the forms of long-distance travelling for religious reasons in the High and Late Middle Ages, crusading has surely captured the imagination of scholars and the general public in the most vivid way. The history

---

[80] Beal (ed.), *Buddhist records of the Western World*, vol. I, 74–75, 89.
[81] Letts (ed.), *The pilgrimage of Arnold von Harff*, 99.

of the Crusades has always been, and still is, a more intensely cultivated subject than the history of medieval pilgrimage. Yet, contrary to conventional wisdom, there is no evidence that the Crusades were a catalyst for the circulation of technical knowledge between Asia and Europe. The only area of technology to which the Crusades probably *did* make a seminal contribution was that of military architecture. Crusaders returning from the Levant brought knowledge about building robust castles of stone to Western Europe, where, especially in the twelfth and thirteenth centuries, this knowledge exerted a strong influence on the construction of noble strongholds and fortifications of cities.[82] There is no doubt that in the High and Late Middle Ages, various items of technical knowledge, for example concerning the use of magnetic needles, the making of Greek fire or the growing of new types of crops (such as sugar), found its way from Asia to Europe by way of the Levant. But there are few indications that crusaders, rather than pilgrims, merchants, seafarers or other categories of travellers to the East, formed a crucial link in the process of circulation. It was not the Crusades that opened the door to the free flow of information between Asia and Europe. Circuits of exchange, as we have seen above, were already in existence before the eleventh century and they comprised more routes than that via the eastern littoral of the Mediterranean.

Comparisons

We can thus find several parallel patterns in the circulation of technical knowledge in China and Europe between the eleventh and sixteenth centuries. Knowledge circulated partly via the spread of drawings and models, and partly through movements of skilled labour and medium- and long-distance travelling for religious reasons. But there were important differences, too. In China, technical information circulated in script and print much earlier than in Europe. Combinations of text and images soon followed. In Europe, by contrast, manuscripts with technical information in verbal and visual form did not begin to spread until the fifteenth century. Diffusion in print commenced about a hundred years later. The 'machine-books' that started to appear in Europe differed from the technical works published in China in a significant respect: they recorded not only technology-in-use, but also projected images of technology that might be developed in future.

[82] Glick c.s. (eds.), *Medieval science*, 348; cf. White, 'Crusades', 285–289, France, 'Technology'.

Moreover, circulation of technical knowledge in China was subject to a greater degree of central regulation than in Europe. From the Song onwards, state agencies took the lead in recording, storing and reproducing technical information. Following the introduction of a system of examinations for state officials, and the regular travelling of scholars that went with it, movements of skilled labour from the Tang onwards, and to a greater extent from the Ming, likewise came under the control of the government to a great extent. Nothing like these government-managed arrangements could be found in Europe. In Europe after c.1000, however, religious institutions, rituals and practices loomed larger in the circulation of technical knowledge than in contemporary China. In Latin Christendom, travelling by monks and pilgrims beyond the immediate surroundings of a village, town or district was a more common occurrence than in the Chinese empire. Formally structured networks between monastic houses that emerged after the tenth century probably favoured the circulation of technical knowledge across great distances, too. From the sixteenth century onwards, patterns of circulation of technical knowledge in China and Europe would diverge steadily further.

Religion and the Circulation of Technical Knowledge After 1500

After catching up with China in the circulation of technical information in script and print, Europe went through a set of further changes that were unmatched in the Ming or Qing Empires. These unprecedented changes did not only relate to the kinds of technical publications and the nature of visual representations of technical knowledge, but also concerned the scope and intensity of movements of people, the manner in which the accumulation and storage of technical knowledge was coordinated and the ways in which technical information was made accessible to different groups of people. This section will show that religious institutions, rituals and practices had in some respects a significant effect on those changes, but that they contributed more to differences *between* China and Europe than to differences *within* Europe. Religious factors were more instrumental in creating the Great Divergence than the Little one.

Texts, Images and Artefacts

Prior to the twentieth century, the ultimate state of perfection in Chinese technical literature was reached in the last decade of the Ming. The *ne plus ultra* in Chinese writing about technology was the *Tiangong kaiwu*

(The exploitation of the works of nature), published in 1637. The *Tiangong kaiwu*, which Needham calls 'China's greatest technological classic',[83] gave an overview of almost all productive technologies that were then in use in China. Its scope ranged from agriculture, textile-making, salt-making, sugar production and ceramics to deep drilling, iron-working, paper-making, shipbuilding, the production of vehicles and the manufacture of weapons. The major technologies that were *not* covered were building, irrigation, printing and the growing and processing of tea.[84]

The author of the *Tiangong kaiwu* was a scholar, Song Yingxing, who, after failing five times to pass the highest level exams in Beijing, served in a low-level position as an education official in Jianxi.[85] Aside from producing a compendium of unusual scope, Song Yingxing also pulled off a remarkable feat by combining quite detailed descriptions of machines, implements and technical processes with more than 120 illustrations, most of them new. Although the descriptions were not invariably complete or consistent, and the texts and drawings did not always fully match (perhaps because Song's publisher also influenced the choice of illustrations), the *Tiangong kaiwu* showed a breadth and depth of information that was not matched by anything that went before it, and the work would remain unsurpassed for over 250 years. Part of this information was borrowed from earlier writings or eyewitness accounts, and part of it derived from Song's own observations during his journeys through the empire. Travelling to exams in Beijing certainly offered plenty of opportunities to take a look around.[86] It is not entirely clear for which audience Song's book was intended. It was perhaps aimed not only at his peers, the educated elite, but also at broader sections of the population, such as merchants, craftsmen and farmers, who, thanks to the sharp drop in book prices in the sixteenth century, may also have been able to buy a copy of this work, and may have consulted it with profit in their own practice. After its first appearance in 1637, Song's book went through a slightly revised, second edition between 1650 and 1680, but it was never completely reprinted

83 Needham, *Science and civilisation*, vol. IV part 2, 171.

84 Schäfer, 'Congruence', 35–60, Golas, '"Like obtaining a great treasure"', 569, Elman, *On their own terms*, 122, Zen-Sun, E-tu and Shiou-Chuan Su (tr.), *Sung Ying-Hsing*, MCullen, 'Interface', 298–300, 314–316. A new, full-scale analysis has recently been published by Schäfer, *Crafting of 10,000 things*.

85 Golas, '"Like obtaining a great treasure"', 572–573.

86 Golas, '" Like obtaining a great treasure"', 574–589, Wagner, 'Song Yingzing's illustrations', 616–621, Bray, 'Introduction: the powers of *tu*', 58–59, 67–68.

again in imperial times.[87] About three-quarters of the original text was incorporated into an encyclopaedia, the *Gujin tushu jicheng* (Complete collection of pictures and books of old and new times), published in 1728 with 64 'officially sanctioned' copies.[88] As a separate entity, Dagmar Schäfer observed, the *Tiangong kaiwu* was probably 'just preserved over a relatively short period of approximately forty years and then totally disappeared until the nineteenth century'.[89]

The *Tiangong kaiwu* was the top of a pyramid. After 1500, government agencies and commercial presses produced numerous other, more specialized, books on technical subjects as well. The 'culture of manuals' for the benefit of scholar-officials reached its peaked in the late Ming and Qing.[90] *Nongshu* continued to appear in great numbers. An estimated 127 agricultural treatises were published under the Ming and another 199 under the Qing.[91] Mechanical devices and implements used in agriculture and other areas of production were described and depicted in agricultural compendia like Xu Guangqi's *Nongzhen quanshu* (1639), in Wang Zheng's *Explanations and diagrams of various machines* (1627) and in sundry encyclopaedias for daily use which poured from commercial presses from the sixteenth century onwards.[92] Great, classic compendia in water management were produced by scholar-officials during the late Ming and early Qing, too. Pan Jixun's *He fang yi lan* (An overview of river defence), composed in 1590, was still regarded as a 'standard guide' in the late eighteenth century. Jin Fu's *Zhi he fang lue* (Methods of river control), presented to the court in 1689 but not printed until 1767, 'long exerted great authority'. Numerous other works on this subject followed in the eighteenth century.[93] And movable-type printing, which was absent in the *Tiangong kaiwu*, was covered in full in works composed by scholar-officials in the eighteenth century.[94]

During the late Ming, for the first time, China also saw the appearance of books on technology that were based on European models. Jesuit

---

[87] Golas, '"Like obtaining a great treasure"', 573, 584–585, 594–595.
[88] Schäfer, 'Congruence', 42.
[89] Schäfer, 'Congruence', 44.
[90] Will, 'Apprentissage', 28–36.
[91] Deng, *Development versus stagnation*, 3–5.
[92] Needham, *Science and civilisation*, vol. IV part 2, 170–171, Engelfriet, *Euclid in China*, 340–341, Bray, 'Agricultural illustrations', 544–545, Elman, *On their own terms*, 110, Bray, 'Introduction: The power of *Tu*', 51–53.
[93] Dodgen, *Controlling the Dragon*, 20, 178, Needham, *Science and civilisation*, vol. IV part 3, 325–326, Bray, 'Introduction: The power of *Tu*', 31–32.
[94] McDermott, *Social history of the Chinese book*, 13.

missionaries in China and their first Chinese converts produced writings that built directly on technological literature published in Europe. Sabatino de Ursis and Xu Guangqi, for example, jointly published *Western techniques of hydraulics* (1612), Johannes Schreck compiled together with Wang Zheng *Diagrams and explanations of the marvellous devices of the Far West* (1627), and Adam Schall von Bell with Chinese assistants made a translation of parts of Georg Agricola's *De re metallica* (1638–1640), which has not been preserved.[95] Many of the illustrations and descriptions contained in the first two works were copied from European books on machines and mechanics published between the fifteenth and early seventeenth centuries, specifically those by Agricola, Jacques Besson, Agostino Ramelli, Simon Stevin, Faustus Verantius, Heinrich Zeising and Vittorio Zonca.[96] Except for the translation of Agricola, which appears to have been lost during the transition to the Qing,[97] none of these works from the last decades of the Ming has since dropped completely out of circulation in China. But their survival mainly took place in an indirect way. Parts of the works produced by the Jesuits and their Chinese collaborators (including many illustrations) were included in dynastic encyclopaedias composed in the eighteenth century. They did not circulate as separate books. The real significance of the appearance of books modelled on European examples thus probably lies more in what it tells us about the relative state of affairs in technological literature in Europe and China than in its actual importance in the circulation of knowledge inside China. In writing on mechanics and machine-building, by the early seventeenth century, Europe had become an example for China instead of the other way around.

According to Sinologists, one aspect of technological literature where, from the sixteenth century, artists and engineers in Europe significantly advanced beyond the level reached in China, was the visual representation of technical knowledge, especially mechanical devices. Whereas technical illustrations in Europe in the fifteenth and early sixteenth centuries steadily became more realist, detailed and precise, and thus became more effective as a means of communicating knowledge, techniques of graphic representation in China did not improve to such a degree that

[95] Jixing, Vogel and Theisen-Vogel, 'Übersetzung', 171, 191–192, Engelfriet, *Euclid in China*, 327–328, Elman, *On their own terms*, 110.

[96] Jäger, 'Buch von den wunderbaren Maschinen', Engelfriet, *Euclid in China*, 338–340, Needham, *Science and Civilisation*, vol. IV part 2, 211–218.

[97] Jixing, Vogel and Vogel, 'Übersetzung', 190.

illustrations could give a realistic picture of complex machinery. Technical illustrations in China remained more simple and crude than those in Europe. Chinese artists or engineers did not show an inclination to 'think' or 'investigate' on paper' either. They did not try out new ideas or imagine new devices by drawing or sketching variations on paper. There was no Chinese Mariano Taccola or Leonardo da Vinci[98]

In the corpus of technical literature published in Ming and early Qing China as a whole, books on agriculture made up a much larger proportion than books on mechanical devices. Chinese *literati* were more concerned with the writing and preservation of agricultural books than with the production and diffusion of works on machine-making.[99] Faced with the continued, massive growth of the population in the eighteenth century, the Qing authorities chose to promote the quantitative expansion of agricultural production by exploiting the full potential of existing techniques, rather than to encourage bold innovations or changes in the organization of the economy.[100]

In Europe, the emphasis in the literature was entirely different. Although writings on farming also appeared in early modern Europe (and still enjoy some status as sources for agricultural history), these publications are, in contrast with China, not considered to have been a relevant factor in the contemporary transmission of knowledge, nor in the continuity of agricultural development as such.[101] More important were *theatri machinarum* or 'machine-books', i.e. books on mechanical devices, illustrated with drawings and other pictures, which became a new genre in the book market in the late fifteenth century. Machine-books were originally mostly composed by engineers from Italy, France and Germany, as a means to advertise their own skills, expertise and ingenuity.[102] The books by Agricola, Besson, Ramelli, Stevin, Verantius, Zeising, Zonca and other authors, whose descriptions and illustrations were borrowed in Sabatino de Ursis and Xu Quangqi's *Western techniques of hydraulics* and Johannes Schreck

---

[98] Bray, 'Introduction: The power of *Tu*', 71–72, Golas, 'Technological illustration', 58, idem, '"Like obtaining a great treasure"', 591–593, McGee, 'Origins', 73–82, Long, 'Francesco di Giorgio and Leonardo', 138–139.

[99] Cf. Deng, *Development and stagnation, passim*, on the role of literati.

[100] Will, 'Développement quantitatif'.

[101] See for example Hoffman, *Growth in a traditional society*, Overton, *Agricultural revolution in England.*

[102] Dohrn-van Rossum, 'Migration-Innovation-Städtenetze', Lindgren, 'Ordnungsprinzipien', 205–214, Long, *Openness*, chapter 4, Lohrmann, 'Das Maschinenbuch', Popplow, 'Why draw pictures', 21–25.

and Wang Zheng's *Diagrams and explanations of the marvelous devices of the Far West*, all formed part of this very tradition.[103]

In the later seventeenth and eighteenth centuries, the machine-books tradition reached its climax with the publication of books on mills and other sorts of mechanical devices in the Dutch Republic. The first 'mill-book', called *Architectura Mechanica Moole boek*, composed by a Swedish millwright, Pieter Linpergh, was published in Amsterdam in 1686. Linpergh's book, reprinted in 1727, was supplanted in the 1730s by two new mill-books composed by millwrights from Holland itself, which surpassed the work of their predecessor in both the quality and quantity of illustrations and descriptions: the *Theatrum machinarum universale, of groot algemeen moolen-boek* by Johannis van Zyl (with engravings by Jan Schenk) and the *Groot volkomen moolen-boek* by Leendert van Natrus and Jacob Polly. In these works, every type of windmill was carefully described and depicted, with measures and all. Engravings printed on folio-size pages showed almost every relevant detail. Mill books circulated among *literati* and entrepreneurs at home and abroad, as well as among millwrights themselves.[104]

Both in China and in Europe, the eighteenth century was the great age of encyclopaedias. It was an era in which Herculean efforts were made to produce all-embracing overviews of the state of human knowledge. But there were some telling differences, too. In sheer scale, the project undertaken on the order of Emperor Qianglong in China dwarfed all ventures carried out in Europe. Between 1772 and 1781, all published books and unpublished manuscripts that could be found in the Qing Empire (some 11,000 titles in all) were sent to an imperial commission in Beijing, which was charged with reviewing, selecting and copying them. These copies were then included in a huge compilation called the *Siku quanshu* (Imperial library in four treasuries). The *Siku quanshu* eventually contained copies of 3,700 works in a set of 36,500 volumes. A complete copy of this set was installed in each of the four principal imperial palaces: two in and near Beijing, one in Chengde and one in Mukden. Another three copies were later housed in cities in the Lower Yangzi region.[105]

103 The machine drawings from the Late Middle Ages to c.1650 contained in these books are now accessible through the Database Machine Drawings, created in Berlin as part of a digital research library, The Archimedes Project; see http://dmd.mpiwg-berlin.mpg.de.

104 Davids, *Rise and decline*, 487.

105 Rowe, *China's last empire*, 88, Guy, *The emperor's four treasuries*, 1, 104, 107, 110.

Among the most massive encyclopaedic projects launched in Europe after 1750 were the *Encyclopédie ou dictionnaire raisonné des sciences, des arts et des metiers*, the *Descriptions des arts et des métiers* and the *Schauplatz der Künste und Handwerke*. The *Encyclopédie*, edited by Denis Diderot and Jean d'Alembert between 1751 and 1782, numbered seventeen volumes with articles and a further thirteen volumes with illustrations, running eventually to 18,000 pages in all. The *Descriptions*, published under the auspices of the Royal Academy of Sciences in Paris between 1761 and 1788, comprised 113 volumes with 13,000 pages and 1,800 illustrations. The *Schauplatz*, compiled by Johann Heinrich Gottlob von Justi and others, consisted of thirteen volumes that appeared between 1761 and 1788. Apart from their size, these European projects also differed from the Chinese version in a number of other significant respects. These publications reached a much broader audience than the *Siku quanshu*. The total number of copies of the *Encyclopédie* alone amounted to 25,000 before 1789.[106] Encyclopaedias in Europe, moreover were not mere sets of copies of existing works. Most of the written contributions and illustrations in these books were entirely new. And as the titles suggest, they consisted, for a large part, of state-of-the-art surveys of technical knowledge. By the middle of the eighteenth century in Europe, the study of technical practices itself had turned into a special branch of knowledge, called 'technology'.[107]

Eighteenth-century Europe also saw the rise of an entirely new genre of technical publications, namely technical journals. Local newspapers and gazetteers, which had spread in China since the Song and circulated in increasing numbers since the late Ming, sometimes contained some information about techniques, such as practices of salt production.[108] The new weekly, monthly or annually published technical journals in Europe were of a different order. These periodicals were specifically intended as a means of information and communication about a state of affairs or a particular technological event, such as the appearance of a new invention or an important publication. Some of these periodicals, such as the *Chemisches Journal für die Freunde der Naturlehre, Arzneygelehrheit, Haushaltungskunst und Manufaktur* were devoted to specific areas of technical knowledge, while others, like the *Transactions of the Society for the encouragement of arts, manufacture and commerce* or the *Intelligenzblätter*, which

---

[106] Darnton, *Business of Enlightenment*, 33–37.

[107] Troitzsch, *Ansätze technologischer Denkens, passim.*

[108] Brook, *Confusions*, 93, 171–172, Gerritsen, 'Fragments of a global past', 138–143, Yoshida, *Salt production techniques*, 6, 8–9.

arose in large numbers in Germany in the late eighteenth century, were more general in nature. A contemporary description of contributions to the *Intelligenzblätter* insisted that they exclusively dealt with 'Die Commercien, die Manufacturen, die Fabriken, das Brauwesen, der Gartenbau, die Landwirthschaft, und andere Stadt- und Landnahrungsgeschäfte'.[109]

Besides technical literature, which codified information in verbal and visual form, in Europe, artefacts became an increasingly important medium for the circulation of knowledge. The use of models was no longer restricted to the world of building and architecture.[110] Scaled-down models of machines, which had existed in Antiquity but had fallen into disuse since, reappeared in Europe in the Late Middle Ages, and from the sixteenth century onwards became an important means of representation in communication between engineers and patrons, in instruction on machine-making and in applications for patents or privileges for technical innovations. Individuals and institutions began to bring together collections of three-dimensional models, which often became accessible to the public at large, too.[111]

By 1800, people in Europe could thus keep themselves informed about the state of the art in technology in a variety of ways. A fine example of a consummate collector of knowledge at this time is that of Maarten Nederdijk, a carpenter and millwright from Rotterdam, the Netherlands. At his death in 1809, Nederdijk possessed a library of some 160 printed books. The assortment of texts in his library ranged from mill books and treatises on architecture to manuals on physics, chemistry and mathematics. Moreover, Nederdijk had also gathered a valuable collection of models, including a small copy of a steam engine.[112]

What was the role of religious institutions, rituals and practices in this expansion of the circulation of technical texts, images and artefacts after 1500? In the world of Latin Christendom after the Reformation, the Catholic Church was a much more stimulating factor in this area than any of the Protestant denominations. A fine example is the use of cathedrals as 'solar observatories'. Many new Catholic churches in Italy and France built in the seventeenth and early eighteenth centuries were laid out in such a

109 Tschopp, 'Popularisierung', 485.
110 Davids, *Rise and decline*, chapter 7.
111 Popplow, *Models of machines*, *passim*.
112 Municipal Archives Rotterdam Ms. 834, probate inventory Maarten Nederdijk, July 1809; unfortunately we do not know the type of steam engine and whether it was working or not.

way as to allow more or less accurate measurements of the movement of the sun's rays along a meridian line on the floor, which provided much-needed data for the exact calculation of the date of Easter. The design of these observatories involved, as a matter of course, an extensive exchange of astronomical, mathematical and technical knowledge, which took place both through travelling of experts, such as Giovanni Cassini, and by means of the circulation of texts and drawings in books, treatises and articles in scholarly journals.[113] Protestant church building never matched this function of the Catholic Church as a hub of technical knowledge.

Of all the religious organizations, the contribution of the Society of Jesus was, yet again, the most prominent. In the seventeenth and eighteenth centuries, Jesuits acted as vital brokers in the circulation of technical knowledge between China and Europe. Intermediation by the Jesuits worked both ways. Apart from introducing European machine-books to a Chinese audience, they also acquainted European publics with technology in China. Descriptions of the manufacture of porcelain in Jingdezhen sent by the Jesuit François d'Entrecolles in 1712 and 1722, for example, greatly aided European states in creating a porcelain industry of their own and provided the basic material for the article on porcelain-making in the *Encyclopédie*.[114] Jesuits contributed to the corpus of machine-books in Europe, too. Kaspar Schott, a pupil and collaborator of Athanasius Kircher who taught mathematics at Jesuit colleges in Sicily and Würzburg between the 1630s and the 1660s,[115] demonstrated in his writings an intimate acquaintance with the latest developments in physics, as well as a keen interest in practical mathematics and the workings of all sorts of machines and mechanical devices. Books like *Mechanica hydraulico-pneumatica* (1657) and *Technica curiosa, sive mirabilis artis* (1664) were intended not only to spread knowledge of the mathematical sciences, but also to make a contribution to the circulation of information on technical inventions, artefacts and curiosities of nature.[116] Athanasius Kircher himself assembled a priceless set of technical objects in his museum in the Jesuit College in Rome.[117] Yet, apart from the efforts by the Society of Jesus and a few writings by the odd vicar or *abbé*, the role of religious actors

113 Heilbron, *Sun in the church*, esp. chapters 2–5.
114 Vissière (eds.), *Lettres édifiantes*, 178–195, Rowe, *China's last empire*, 84.
115 Hellyer, *Catholic physics*, 123–124.
116 Schott, *Mechanica hydraulico-pneumantica* (Würzburg 1657), idem, *Technica curiosa, sive mirabilis artis* (Würzburg 1664).
117 Findlen, 'Scientific spectacle'.

in the process of circulation of texts, images and objects was decidedly a minor one. In this respect, religious contexts counted for little in China and Europe; but it was different with regard to the movement of people.

Movements of People

While after 1500, movements of people in China continued to be subject to a greater degree of state regulation than in Europe, the state-governed systems of mobility that had been in place during the early Ming showed starkly divergent developments thereafter. On the one hand, the civil examinations system, along with the movements of candidates to regional capitals and to Beijing that went with it as a matter of course, persisted as a fixture of the Chinese state throughout the Ming and Qing periods. The Qing emperors also strictly enforced a system of rotation of officials, which entailed that officials at a provincial level served no longer than three years in the same place and those at a local level, no longer than half that time.[118] Scholar-officials thus continued to move around on a regular basis after they had earned their degree. The travelling of *literati* for the purpose of examinations or job rotation may have facilitated the circulation of knowledge, although it is impossible to offer solid proof of this, because written accounts of such journeys are exceedingly rare.[119] The corvée system, on the other hand, had already begun to lose its edge in the second half of the fifteenth century and it disappeared entirely, as far as craftsmen were concerned, in the mid-sixteenth century. From 1562, shift artisans could buy off their labour service with a money payment.[120] The system was not restored under the Qing.[121] When the system of obligatory labour collapsed, the regular pattern of travelling by craftsmen to the capital must have lost its rationale as well. This, in turn, reduced the opportunity for the circulation of technical knowledge by movements of people.

Guilds formed another potential channel for the circulation of knowledge and skills across the empire. In contrast with Europe, however, guild-related circulation did not hinge upon the existence of a tramping system (which was absent in Late Imperial China), but on the peculiar composition of guilds. Chinese guilds were often composed of people from the

118 Rowe, *China's last empire*, 39, 45–48.
119 Eggert, *Vom Sinn des Reisens*, 53.
120 Friese, *Dienstleistungs-System*, 128–129.
121 Moll-Murata, *State and crafts*, 51.

same geographic area. Scholars have highlighted this characteristic as a striking difference between Chinese and European guilds.[122] Peter Golas mentions the case of the Leather Box guild in Beijing, founded in 1689, which as late as the 1940s still consisted exclusively of people originating from Shandong.[123] The principle of common geographical origin could imply that knowledge and skills in specific occupations travelled a very long way. However, common geographical origin appears to have become *less* important as a basis for organization from the eighteenth century onwards.[124] As a consequence, knowledge and skills in guild-organized crafts likely moved across shorter distances than before.

The demise of the corvée system and the gradual decline of the principle of common geographical origins in guilds did not imply that long-distance circulation of technical knowledge through movements of skilled people entirely came to a halt. Circulation of technical information could still occur on an ad-hoc basis or take place outside formal institutional structures. In the eighteenth century, for example, specialists in reverse glass painting from Guangzhou were frequently summoned to Beijing to work in the imperial palace workshop, where they sometimes came into contact with resident painter-officials.[125] Another mode of circulation is exemplified by movements of Chinese sugar-mill workers to the environs of Dutch Batavia, Java, during the late seventeenth and early eighteenth centuries. In about 1730, the area around Batavia boasted some 130 sugar plantations, almost all of them owned and managed by Chinese.[126] The technology employed for processing sugar cane on these plantations, including the use of animal-driven double-roller mills, was also imported from China.[127] The sugar mills were probably built by machine-makers who, like most Chinese living at Batavia,[128] hailed from Fujian. Neither in Fujian nor in Batavia, however, was there any formal institutional structure in place to keep this migration going for an extended period of time.

---

[122] Golas, 'Early Ch' ing guilds', 563, Moll-Murata, 'Chinese guilds', 220, 240.
[123] Golas, 'Early Ch' ing guilds', 564.
[124] Golas, 'Early Ch' ing guilds', 558, 564–565, Rowe, *Hankow*, 252–288.
[125] Curtis, *Glass exchange*, 51–52.
[126] Needham, *Science and civilisation*, vol. VI part 3, 438, Blussé, *Strange company*, 90–91.
[127] Needham, *Science and civilisation*, vol. VI part 3, 438, Leidelmeier, *Van suikermolen tot grootbedrijf*, 18–21.
[128] Blussé, *Strange company*, 147.

Inside China, people also travelled, of course, for reasons other than official assignments. Private travel vastly increased in the late Ming. While state travel services in this period contracted due to lack of funds, commercial services catering for the needs of travellers expanded. The experiences of these travellers sometimes found their way into script or print.[129] Accounts of trips undertaken for private pleasure, edification or business affairs do not betray any interest in technical subjects, however. They were not technological journeys in any meaningful sense of the word. The private trips of which descriptions have been preserved were bound for lakes, rivers, waterfalls or mountains (sometimes for religious reasons), but seldom for cities and never for sites of production.[130] Nonetheless, trips that were recorded in travel accounts may still have made a contribution to the circulation of knowledge and skills.

Religious institutions, rituals and practices may have played a part in this, too. Susan Naquin has suggested that temples served, among other things, as 'living examples of special technologies and arts'. Temples in bigger cities offered models of building techniques, materials and decoration for craftsmen from smaller towns.[131] Religious festivities, such as 'silkworm festivals' held in temples for local deities in eighteenth-century Jiangnan, moreover provided excellent opportunities for the spread of new techniques in sericulture. These festivals attracted visitors from many different localities in the surrounding countryside.[132]

From the great upheavals of the sixteenth century onwards, Europe experienced much higher levels of religious diversity than in the past. In the previous chapter, I examined the impact of these religious variations after 1500 on the formation of human capital through formal and informal learning. How, and to what extent, did religious transformations also have an influence on patterns of mobility of people? If patterns of mobility of people changed, this may have affected the circulation of technical knowledge and skills as well. To answer this question, I will look at the mobility of Europeans from various perspectives: the size and duration of movements, the categories of labour involved, the distances covered and the nature of the journeys performed.

---

129 Brook, *Confusions of pleasure*, 173–182, Eggert, *Vom Sinn des Reisens*, 23–32, 54–57.

130 Eggert, *Vom Sinn des Reisens*, 57–68.

131 Naquin, 'Temples', 199–207.

132 Cho, 'Sericulture songs'. Religious cults also existed at salt production sites. Hans Ulrich Vogel has interpreted these as a kind of economic exchange: in return for temple building, worshiping and sacrifices, participants expected a high output of salt or brine. Vogel, 'Nature in traditional Chinese salt production', 501–502.

Heinz Schilling has coined the term 'confessional migration' to describe migration that was linked to the process of 'confessionalization' – a combination of religious reform, state-building and social discipline practised in many European countries from the middle of the sixteenth century onwards. The adjective 'confessional' in combination with 'migration' primarily refers not to the motives of migrants, but to the impact that this process of confessionalization had on the setting in which migration processes took place.[133] Stock examples of the relevance of confessional migration for the circulation of technical knowledge are the huge wave of Calvinist refugees from the Hapsburg Netherlands between c.1560 and 1600 and the massive exodus of Huguenots from France at the end of the seventeenth century.

Historians of these large swells of Protestant migration have been keen to stress their seminal importance not only for the political and cultural history of the sender and receiver countries, but also for the history of economic development and the diffusion of technical knowledge. In his major study on the consequences of the Revocation of the Edict of Nantes, for example, Warren Scoville has claimed that 'the evidence that Huguenots helped diffuse many of France's industrial secrets and processes in England, Ireland, Holland, Germany, and Switzerland [was] convincing'.[134] The physical displacement of tens of thousands of craftsmen, merchants and manufacturers, induced by confessionalization, doubtless implied a massive movement of skills. In this way, religious transformations led to an acceleration and temporary expansion of the circulation of technical knowledge. In various sectors of industry, confessional emigration from established sites of production resulted in the emergence of new centres of expertise, such as Amsterdam in sugar refining or London in silk-weaving, which in time became hubs in the circulation of knowledge in their own right.[135]

The reverse side of confessionalization was that host countries could be reluctant to allow migrants of a different creed from that of the locally dominant confession to become fully integrated in their new environment. Migrants of a different creed may also have themselves preferred

---

[133] Schilling, 'Confessional migration and social change', 322.

[134] Scoville, *Persecution of the Huguenots*, 443, Scoville, 'Minority migration', 347–360, Scoville, 'The Huguenots and the diffusion of technology', 294–311, 392–411. See also e.g. Briels, *Zuid-Nederlanders in de Republiek*.

[135] Liu, *Immigrants and the industries of London*, 175–218, Poelwijk, *'In dienste vant suyckerbacken'*, chapter 3.

to keep social contact with the host community to a minimum. Limited tolerance and economic discrimination in Cologne and Aix-la-Chapelle from the early seventeenth century onwards, for example, induced many Protestant refugees to leave these Catholic Imperial cities for Frankfurt, for smaller towns in the Rhineland or for places in the countryside.[136] In late sixteenth-century London, 'aliens' (at that time mostly Calvinists) were only permitted to stay in the city if they taught their arts to Englishmen and, as a survey from the early 1590s showed, a large number of them indeed went on to employ English servants, journeymen or apprentices in their households. The degree of social mixing between strangers and Londoners remained low, however, as children of alien-born parents were not allowed to serve as apprentices to native masters and immigrants largely kept to their own churches and communities. Faced with growing hostility from native artisans and merchants and increasing discrimination on the part of the government authorities in the 1590s, many Calvinist families decided to leave London for the Dutch Republic, which by then appeared to offer much better economic and political opportunities.[137] The actual impact of confessional migration on the circulation of technical knowledge was thus heavily dependent on the nature and extent of confessionalization in particular local or regional settings.

Great waves of confessional migration, moreover, might easily obscure the importance of other migration flows, whether inspired by religious motives or not. Long before the religious upheavals of the sixteenth century, migration flows could contribute to the emergence of new centres of production in particular branches of industry. 'Alien' brewers (mostly from Holland), for example, started the production of hopped beer in London in the fifteenth century.[138] And religious transformations in the late sixteenth century did not impede the long-distance circulation of technical knowledge by small-group migrations between Protestant and Catholic regions. Craftsmen from Catholic Venice, lured by Dutch recruiting agents, helped to establish the glass-making industry in the United Provinces in the early seventeenth century. Immigration of experts from the Prince-Bishopric of Liège contributed to its survival.[139]

---

136 Schilling, 'Confessional migration', 328–329.

137 Liu, *Immigrants*, 126–130, 146–167.

138 Liu, *Immigrants*, 264–266.

139 Klein, 'Nederlandse glasmakerijen', 35–36; Venetian glassmakers continued to emigrate despite legal penalties on the export of technology. See Gasparetto, *Vetro di Murano*, 70, 110–113, Ashtor, 'Factors', 20–21.

Movements of skilled people, however, did not only take place – to borrow an old Braudelian turn of phrase – in the 'time of events'; they also showed structural patterns. Great waves of migration or sequences of small-group migrations occurred against the background of continued movements of individuals between cities and regions in search of training or employment. Tramping of journeymen was a widespread phenomenon in Central Europe throughout the early modern period and existed in eighteenth-century France and, to a lesser extent, in England as well.[140] In the Dutch Republic, too, at least from the late seventeenth century onwards, a substantial proportion of journeymen regularly travelled from one city to another, even if an organized system of *Wanderschaft* was absent.[141]

Historians have underscored the importance of this regular *Wanderschaft* or *compagnonnage* of journeymen as a medium of technology transfer, even though the principal driving force of the system was not technological but economic.[142] Larry Epstein, in particular, has explained its relevance with the following argument: (1) only migration by trained craftsmen 'is likely to have had a significant impact' on 'technical diffusion' in Europe in the Late Middle Ages and the Early Modern Period, given the fact that 'mass schooling' and 'the harnessing of science to industry were unknown' and 'most technical knowledge was tacit and embodied in its practitioners'; (2) transitory migration in the form of journeymen tramping was, in this regard, more important than permanent migration by master artisans; (3) informal tramping was numerically much more significant than tramping under the aegis of formally organized journeymen's associations; and (4) this regular pattern of journeymen travelling was instrumental in the formation of regional 'pools' or 'clusters' with shared technical knowledge and skills, which did not extend across Europe as a whole but largely coincided with 'areas that were institutionally and culturally more homogeneous', such as the German lands.[143]

Were these patterns of artisan mobility affected by the religious upheavals from the sixteenth century onwards? The effect of religious transformations, I would suggest, was more manifest in the case of masters than of journeymen. Confessional affiliation became a relevant factor in the

---

[140] Ehmer, 'Artisans, journeymen, guilds and labor markets', 60–63, Reith, 'Circulation of skilled labour', 117–130, esp. 125.

[141] Davids, 'Guilds'.

[142] Reith, 'Arbeitsmigration und Technologietransfer', 52–53, Epstein, 'Journeymen mobility', Epstein, 'Labour mobility'.

[143] Epstein, 'Journeymen mobility', especially 411, 413, idem, 'Labour mobility', especially 250, 252, 266.

admission of a craftsman as a master, citizen and guild member (which were normally interlinked categories). In Central Europe in the seventeenth and eighteenth centuries, Catholics were by and large barred from gaining citizenship and/or guild membership in cities like Berlin, Bremen and Frankfurt, where Protestant confessions reigned supreme. This meant that they were also deprived of the opportunity to become masters. A similar disadvantage existed for Protestants in Vienna, where Catholicism was the dominant confession. Still, even in such 'mono-confessional cities', exceptions could be made for immigrants of a different faith on an individual basis, provided they brought special, highly valuable skills.[144] The situation in France by and large resembled that in 'mono-confessional' cities in the German lands: adherence to the dominant church (the Catholic Church) was normally a precondition for admission to the status of master, with the proviso that the royal government could grant exemptions to foreign Protestants commanding ample supplies of capital or special skills.[145]

The United Provinces, where the dominant, 'public' church was Calvinist, presented a more complex picture. From the late sixteenth century, the inland provinces of the Republic – to the east of the city of Utrecht – conformed to the 'German' model, setting up high barriers for admission to citizenship and guild membership for Catholics, Lutherans, Mennonites and other 'dissident' Christian groups. In Holland, by contrast, no such policies existed – except with regard to Jews. There were variations in the share of different confessions by craft or trade, to be sure. Lutherans from Germany, for example, were overrepresented among Amsterdam's bakers and tailors, but underrepresented among the city's goldsmiths. But such variations sprung from informal selection mechanisms, from differences in formal education or from disparities in economic resources (e.g. the amount of capital needed to set up shop) rather than from any formal regulations that discriminated against members of particular confessions.[146]

As the status of master in a particular craft or trade became more associated with a specific confessional background, the circulation of knowledge in crafts and trades within cities may have become more specialized along religious lines, too. However, before jumping to conclusions about

144 Pallach, 'Fonctions', 374–375, Steidl, *Auf nach Wien!*, 59–60.

145 Pallach, 'Fonctions', 376, Kaplan, *Divided by faith*, 248.

146 Lourens and Lucassen, '"Zunftlandschaften" in den Niederlanden und im benachbarten Deutschland', 18–19, Prak, 'The politics of intolerance', 159–175, Kuijpers, *Migrantenstad*, 130–135, 217–219, 283–284.

a kind of 'Weberian' path dependency in technology, we should realize, of course, that a bias in one direction in City A (say, the goldsmith's trade becoming a specialty of Lutherans) may have been offset by a bias in another direction in City B (the goldsmith's trade becoming dominated by Catholics), the net result being that seen across a wider area (for example, France, or the Holy Roman Empire), the association between confessions and technical specialties was much weaker. Disparities of knowledge between cities may also have been evened out by the continued interurban movements of journeymen, which after all were much less affected by religious transformations than the selection of masters. Moreover, a bias at a local level against people of a particular confession could ultimately be corrected by political means. The rise of the share of Catholics in almost every craft and trade in Colmar from the end of the seventeenth century had much to do with the transfer of the city to French rule in 1679, and the subsequent influx of Catholics from the countryside. In eighteenth-century Vienna, it was the Imperial Government that began to weaken the grip of the guilds on admission policies, thereby helping to improve the chances of outsiders from different religious backgrounds.[147]

As religious dividing lines hardened in the late sixteenth century, chances for admission to the status of master in many parts of Europe thus became increasingly dependent on site-specific confessional relations. Technical specializations at the level of master-artisans could therefore become biased by confession as well. However, this effect could be mitigated by several countervailing forces.

Barriers to the circulation of technical knowledge and skills at the level of journeymen were lower than those facing masters. In the Holy Roman Empire, during their *Wanderschaft* years, journeymen continued to travel relatively easily across borders between regions adhering to different faiths. Catholic and Protestant journeymen continued to work more or less harmoniously side by side.[148] Tramping patterns coincided, to a lesser extent, with 'areas that were institutionally and culturally more homogeneous' than Epstein assumed. Mid-eighteenth century data from (Catholic) Lower Austria, for instance, show that sock makers, although they 'were not supposed to travel beyond the immediate hinterland', still wandered frequently to Italy and France, followed by (Protestant) Saxony and Prussia, and that glove makers preferred to tramp to France and

[147] Wallace, *Communities and conflict*, chapter 8, Steidl, *Auf nach Wien!*, 60, 85–88.
[148] Pallach, 'Fonctions', 375.

Switzerland.[149] Spatial movement of technical knowledge among journeymen was thus apparently not hampered by religious boundaries in any serious way.

Some regions that were long-standing pools of skilled migrant labour persisted in sending both masters and journeymen to other parts of Europe, regardless of their religious complexion. After the Thirty Years War, hundreds of masters and journeymen from Tyrol who were renowned for their skills in the building trades would travel each year not only to work at Baroque churches, palaces or monasteries at other places in Central Europe, but also to hire themselves out on different construction projects in the Low Countries, France and Protestant territories in Germany.[150]

Religious boundaries between territories in Central and Western Europe were further circumvented by a practice called *Auslauf*. *Auslauf* meant that people from a 'dissident' confession in a particular city or region regularly crossed the border to a neighbouring territory where they could worship in relative freedom. Among Catholics, such movements also could take the form of a pilgrimage to a holy site in a nearby country. In the later seventeenth and eighteenth centuries, for example, thousands of Catholics from Holland travelled across the border of the Dutch Republic every year, to visit the Marian shrine in Kevelaer in Upper Gueldres.[151] The growth of such practices of religious travelling meant that co-religionists from territories of a different confessional stamp could keep contact with each other. It implied, too, that artisans, living in different territories but adhering to the same faith, nevertheless had regular opportunities for communication.

Religious transformations in Europe from the sixteenth century had a profound effect on long-distance travel. Religious changes after 1500 contributed to a huge increase in long-distance travel by Europeans, which in turn led to a massive expansion and acceleration in the circulation of technical knowledge and skills. Long-distance travelling by Europeans became much greater in scale and scope than travel by Chinese in the Late Ming and early Qing period. Even though Ming China saw the growth of 'a widespread interest in the novel, the curious and above all the foreign and imported',[152] the emergence of this 'culture of curiosity'

149 Reith, 'Circulation of skilled labour', 140.
150 Reith, 'Circulation of skilled labour', 125–126.
151 Kaplan, *Divided by faith*, 161–171, Wingens, *Over de grens*, 157–174.
152 Clunas, *Pictures and visuality*, 173, Waley-Cohen, *Sextants of Beijing*, 58–59.

did not result in a vast increase in travelling abroad,[153] nor in a significant expansion of written or published reports on other areas of the world. But this was exactly what happened in Europe, and religious institutions played an important part in this process.

The crucial change was the immense expansion of missionary activity outside Europe from the sixteenth century onwards. This outburst of missionary fervour formed part and parcel of the sweeping, comprehensive movement of Catholic Renewal. European missionary organizations built networks than spanned a large part of the known world. In Asia, these networks not only embraced key positions and lines of communication in the Portuguese and Spanish areas of operations (such as Goa, Macao, Diu, Chaul, Damão, Bassein, the Lesser Sunda Islands and the Philippines), but also extended to countless places, regions and routes in East, Southeast and South Asia, where the Iberian powers never could claim any sort of political, military or economic control at all, or where they had lost whatever control they may have once had to Asian rulers or European trading companies before the end of the seventeenth century. The latter category included, among others, China, Japan, Vietnam, Siam, Malacca, Pegu, Bengal, Sri Lanka, and Malabar, as well as Mesopotamia and Persia. All the major religious organizations that engaged in activities across the ocean set up a system of regular reporting on the state of affairs of their missionary efforts by region, and the conditions and contexts in which these operations were carried out. Frequently, they also took care that at least some of these records also reached a wider audience. Franciscans, Dominicans, Augustinians, the *Propaganda Fide*, the *Mission étrangères* and especially the Society of Jesus were all very active in this respect.[154] The range of action of these organizations in Asia and the diversity in geographic origin of their members in Europe ensured that they had access to a wide spectrum of information and could communicate their experiences, observations and findings to a variety of publics at home.

Among the information gathered and recorded by missionaries in Asia were data on the natural environment, as well as things worth knowing on cultures, economies and societies. This constant, purposeful collection of 'useful' knowledge can partly be explained by the fact, as Steven Harris has argued,[155] that religious organizations operating far from their base in

---

[153] Waley-Cohen, *Sextants of Beijing*, 122.
[154] Lach and Van Kley, *Asia in the making of Europe*, vol. III, book one, 301–305.
[155] Harris, 'Long-distance corporations', especially 285–300.

Europe had to devise ways 'to keep their members healthy' and therefore sought, among other things, to gather various sorts of natural knowledge. But religious organizations were also interested in gathering information more generally, because it might help in better realizing the very purpose of their presence in Asia, namely: the propagation of the faith.

The most extensive, systematic surveys of 'useful knowledge' in Asia from the seventeenth century could be found in publications by Jesuit missionaries. The works by Nicholas Trigault, João Rodrigues and Alvaro Semedo, which were based both on their own observations and a massive amount of information gathered by numerous other priests and brothers, presented not just surveys of the history, politics and society of these Asian empires (and the vicissitudes of the Jesuit missions), but also included lengthy discussions on such subjects as husbandry and material culture, and presented generally laudatory accounts of the state of the 'liberal and mechanical arts'.[156] And sometimes letters sent by missionaries in faraway places contained quite detailed information on particular technologies. D'Entrecolles' report on the making of porcelain in Jingdezhen, cited above, is a sterling example of such a specific contribution by missionaries to the circulation of technical knowledge.

The sheer range, coherence and detail in works composed by members of Catholic missionary organizations before the eighteenth century were unmatched by any of the rare surveys of Asian countries produced by Protestant ministers active in territories controlled by European trading companies. The description of Malabar and Coromandel published in 1672 by Reverend Philippus Baldaeus, for example, who in the 1650s and 1660s worked in Dutch East India Company (VOC) settlements in South India and on Sri Lanka, in fact, heavily depended on books written by Jesuits.[157] For a very long time, neither did the Dutch and English trading companies contribute anything like that which had been produced by the religious organizations to the circulation of technical knowledge. *Pace* Harold Cook,[158] before the eighteenth century, values of commerce were less powerful driving forces in the collection of empirical knowledge about the outer-European world than values of religion. All those Jesuit or Franciscan missionaries overseas collecting specimens, drawing maps and

---

[156] Ricci and Trigault, *Histoire de l'expedition chrétienne*, 76–149, Cooper (ed.), *João Rodrigues's account*, chapters 1, 2, 7 and 8, Semedo, *Relazione*, 5–74.

[157] Lach and Van Kley, *Asia in the making of Europe*, vol. III, book two, 911–912, 1063–1064, Baldaeus, *Naauwkeurige beschryvinge van Malabar en Choromandel*.

[158] Cook, *Matters of exchange*, esp. chapters 1 and 11.

writing books and reports were not motivated by the prospect of improving their own income or wealth, or that of the organization to which they belonged. Spiritual autobiographies and questionnaires composed by Jesuits in the sixteenth century studied by Luke Clossey and others rather suggest that the desire for salvation must have been a strong incentive.[159] In their view, labour 'in the vineyard of the Lord' represented a higher reward than earthly riches.

Knowledge and skills also flowed the other way.[160] Jesuit missionaries, as we have seen, brought machine-books from Europe to China. These works formed the basis for books on machines in Chinese by Sabatino de Ursis, Xu Guangqi, Johannes Schreck, Wang Zheng and Adam Schall von Bell published between 1610 and 1640. At the court of the Qing, Jesuits were not only highly valued for their superior knowledge of astronomy and mathematics, but also for their expertise in a wide range of other fields. The skills of Jesuits were at a premium in many different specialties. Jesuit missionaries supervised the founding of cannon, they conducted surveys of the empire, they produced maps, they constructed clocks, automata and scientific instruments, they built machines, they introduced new plants and medicines, they designed fountains, gardens and palaces, they led operations in glass-making and they painted portraits of emperors and scenes of imperial victories.[161] The achievements of Jesuits as technical specialists in China reached their peak during the reign of Emperor Qianlong (1736–1795), when they were forbidden to carry out missionary activities. 'Depuis que les Missionaires sont établis ici', a Jesuit wryly remarked in a letter from Beijing in 1754, 'il n'y a eu aucun empereur qui ait plus profité de leurs services que l'Empereur regnant; et il n'y en a eu aucun qui les ait plus maltraités, et qui ai porté de plus foudroyants arrêts contre la sainte religion qu'ils professent'.[162]

In Europe, meanwhile, as in Late Ming China, long-distance travelling for other than religious reasons perceptibly increased in the early modern period. Private travelling for the purpose of business, pleasure or education became a more widespread practice than before. With the rise of travelling went the growth of a didactic literature on the art of travel. The *ars apodemica*, which flourished in Europe between the mid-sixteenth and

---

[159] Clossey, *Salvation and globalization*, 114–135.

[160] An aspect neglected by Hobson, *Eastern origins*, 200.

[161] Waley-Cohen, *Sextants of Beijing*, 105–121, Curtis, *Glass exchange*, 36–37, 41–53, 60–61, 113–117, Musillo, 'Reconciling two careers'.

[162] Vissière (eds.), *Lettres édifiantes*, 455.

late eighteenth centuries, taught people how they should plan their journeys, how they should reflect on their experiences and how they should keep a systematic record of their observations.[163] This didactic literature had no equal in China, although practical guides for travellers and models of travel accounts were not unknown.[164] Like the travel books in China, however, the *ars apodemica* did not offer were instructions concerning the gathering, recording and storing of information on technology. The art of travel remained separate from the art of technical writing. Among engineers, however, from the sixteenth century, it was not unusual to keep notebooks with sketches of remarkable devices that they had observed while travelling. Added to measurements, gear ratios and other useful pieces of information, such sketches served as kind of 'private archive' for the engineers' practice.[165]

From the mid-seventeenth century, finally, a new genre of travelling emerged: travels made with the express purpose of observing technological practices and artefacts. These observations were usually put on record in manuscripts or printed accounts. Technological journeys were frequently undertaken under the auspices of some state agency, according to a set of official instructions. Initially, the favourite destination of technological journeys in Europe was the Dutch Republic. Frenchmen, Englishmen, Scotsmen, Prussians, Italians, Swedes, Danes and Russians, equipped with pencils and notebooks, descended on industrial mills and shipyards in the Zaanstreek, cloth manufactories in Leiden, bleaching fields near Haarlem, ceramic workshops in Delft, pipe-making shops in Gouda, chemical laboratories in Amsterdam, or locks, sluices and bridges anywhere in Holland to grasp the secret of Dutch success.[166] In the second half of the eighteenth century, the focus of attention of technological travellers increasingly shifted to England and France.

Collection, Storage and Accessibility of Information

Besides differences in the nature of data carriers and in the scope and intensity of movements of people, after 1500, variations between China and Europe also emerged in the ways in which technical information was

---

163 Stagl, *Geschichte der Neugier*, Rubiés, 'Travel writing as a genre', Rubiés, 'Travel writing and ethnography', Rubiés, 'Instructions for travellers', Boedeker, '"Sehen, horen, sammeln und schreiben"', Schulz-Forberg, 'European travel and travel writing'.

164 Eggert, *Vom Sinn des Reisens*, 94–110, Brook, *Confusions of pleasure*, 179–182.

165 Popplow, 'Why draw pictures', 36–42.

166 Davids, *Rise and decline*, chapter 5.

collected, stored and made accessible. In Europe, the network of 'fixed places' where information was assembled and preserved assumed a different structure from that which existed in China, and the ease with which the information in those places could be accessed by various groups of people also started to diverge. In Europe, 'establishing', 'controlling', 'selling' and 'acquiring' knowledge – to borrow Peter Burke's terminology – began to look more and more different from their Chinese equivalents.

From the Song onwards, China saw the rise of a substantial body of technical literature. Technical knowledge in the form of manuscripts or printed books was stored in various places. Repositories of technical knowledge in late Ming and early Qing China were located in imperial palaces and central government institutions, in regional administrative centres, in government schools and in private libraries in various places in the empire. The largest collection of books in the empire was housed in the Imperial Library in the capital, which had been rebuilt and replenished after every change of dynasty since the Han period.[167] 'A government's ability to collect books, correct texts, and pass judgement on them had become a sign of its power, and a regime which failed to do this was somehow inadequate', Kent Guy observed.[168]

The first Song emperors commissioned the printing of four huge encyclopaedias.[169] During the early Ming, the Yongle emperor (1402–1424) ordered the compilation of an encyclopaedia that would contain 'all the books of the empire'. The only copy of this encyclopaedia, the *Yongle dadian*, completed in 1409, was housed in the Hall of Literary Profundity at the Imperial Court. Another copy was later stored in the Imperial Archives.[170] The project of collecting, collating, copying and cataloguing all books in the empire carried out under the aegis of Emperor Qianglong in the 1770s and 1780s may have been unprecedented in scale, but was evidently not entirely novel in its kind. The compilation of the Four Treasuries was in fact 'a revival of the long tradition of centralized imperial book collecting', which made eminent sense from the point of view of legitimation of the Manchu regime.[171]

Central government institutions and regional centres of administration had collections of books, too. Under the Song and Ming, for example, sig-

167 McDermott, *Social history*, 49–50, Guy, *Emperor's Four Treasuries*, 13–16.
168 Guy, *Emperor's Four Treasuries*, 12–13.
169 Egan, 'Grains of sand', 35.
170 McDermott, *Social history*, 127–128, Guy, *Emperor's Four Treasuries*, 15.
171 Guy, *Emperor's Four Treasuries*, 16, 34.

nificant holdings could be found in the Hall of Literary Profundity, the Historiographical Institute and the Directorate of Education.[172] In Ming and Qing times, the central agency to coordinate efforts to control the Yellow River, the Office of Rivers and Canals, doubtless possessed copies of, for instance, Pan Jixun's *He fang yi lan* (An overview of river defence) and Jin Fu's *Zhi he fang lue* (Methods of river control). In the late eighteenth century, provincial libraries were one of the sources for the Four Treasuries project.[173]

School libraries were founded all over China under the early Ming. During the reigns of Emperors Hongwu (1369–1398) and Yongle, all state schools in the empire, down to the county level, were provided with a standardized collection of books consisting of Confucian classics, didactic texts and all kinds of reference works, printed and distributed by the state.[174] Under the later Ming, these government libraries were sometimes partly funded by merchants, because the official budgets no longer sufficed. The numbers of books kept in prefectural or county libraries at that time was not very substantial, though.[175] Schools and academies were important markets for booksellers under the Qing. Bookstore managers of Sibao publishers were keen to get orders for textbooks before the beginning of a new school year.[176]

Private libraries began to multiply under the Song,[177] but they only became of increasing significance from the sixteenth century onwards. The size of these libraries increased by leaps and bounds, as the supply of books in print increased. Many of the largest collections in the empire were formed by private owners. At least 40% of these larger private collections under the Ming could be found in the Yangzi Delta.[178] Private collectors in the Late Imperial period mainly came from the ranks of scholar-officials, gentry and merchants;[179] at this time, religious institutions like temples or monasteries were not known for having well-stocked libraries. When in 1772, Emperor Qianlong ordered a survey of all the extant titles of books

172 McDermott, *Social history*, 127–132, Brook, *Confusions of pleasure*, 276 n34.

173 Guy, *Emperor's Four Treasuries*, 89–91.

174 Brook, 'Edifying knowledge', 104–105.

175 Brook, *Confusions of pleasure*, 92–93, McDermott, *Social history*, 133, Lui, *Hanlin Academy*, 20.

176 Brokaw, *Commerce in culture*, 254–255.

177 McDermott, 'Book collecting', 65, 69–70.

178 McDermott, *Social history*, 134, 145–146, Brook, *Confusions of pleasure*, 132, 169.

179 Guy, *Emperor's Four Treasuries*, 46–47, McDermott, *Social history*, 135–138, Brook, *Confusions of pleasure*, 132, 169–171.

in the empire for the benefit of the *Siku quanshu*, nearly a third of the 11,000 titles were provided by private collections.[180] Evidently, government libraries did not possess a copy of every book published in China.

If there was a weak point in the collection and storage of information in books in China, it lay not in the acquisition and preservation of books per se, but rather in their accessibility. Collecting books was not the same as circulating them. The *Yongle dadian*, the greatest hoard of book knowledge amassed before the end of the eighteenth century, could not be consulted by anyone other than the emperor himself, except by special permission (which never seems to have been granted).[181] Of the *Siku quanshu*, no more than seven manuscript copies were produced, which could only be accessed in a few select places. Libraries in central government institutions in the capital were open to scholar-officials and members of the imperial household, but not to the public at large. Except in government schools, institutional libraries open to a wider audience were extremely rare in the rest of the empire. Lending systems for books hardly developed. Private book collections often remained closed to any interested readers other than the owners and their kin. 'The concept of a "community of learners" described more an ideal than the reality of the world of learning in China between 1000 and 1700', Joseph McDermott concluded. The great book collections in the Yangzi Delta 'would have resembled nothing less than virtually locked vaults of knowledge'.[182] And even though books became more numerous from the sixteenth century, Timothy Brook reminds us, 'knowledge still did not circulate without restriction. Craft knowledge tended to be protected as the preserve of specialists or known only locally'.[183] Technological literature that circulated more widely thus did not cover the entire body of locally available knowledge.

Although technological literature in early modern Europe did not encompass all locally available knowledge either, the network of institutes in which it was collected and stored and the ways it was made accessible from the fifteenth century began to show a distinctly different pattern from that which emerged in China. The structure of institutions in Europe was less centralized than in the Chinese Empire, they were more open to a wider public and religious organizations continued to be a major fac-

---

[180] Guy, *Emperor's Four Treasuries*, 90.
[181] McDermott, *Social history*, 127–128.
[182] McDermott, *Social history*, 129–147.
[183] Brook, *Confusions of pleasure*, 132–133.

tor in the collection, storage and dissemination of information in books throughout the early modern period.

Libraries proliferated in Europe from the Late Middle Ages onwards. Fixed sites for the collection and storage of books could increasingly be found not only in princely palaces – imperial, papal, royal or otherwise – cathedrals, monastic houses and synagogues, but also in universities and other educational establishments, in municipal buildings, meeting places of the professions and the homes of private citizens. Until the fifteenth century, the libraries of religious institutions remained the dominant model for all institutional libraries in Europe. Access to collections was restricted. Monastic houses, for instance, seldom lent books to people other than members of their own communities. Libraries normally only contained books in Latin, with a heavy emphasis on religious and philosophical texts. The collection of Emperor Frederick II, probably housed in Palermo, was unique in the thirteenth century in that it was multilingual and truly ecumenic in its range of subjects. Texts in the vernacular, which began to spread from c.1300 onwards, initially figured almost exclusively in the collections of private citizens, such as humanists or merchants.[184] As writings on technical subjects before 1400 were rare, however, it would be going too far to regard either institutional or private libraries in this period as a kind of hub for the collection, storage or dissemination of *technical* knowledge.

However, this began to change from the fifteenth century onwards. Access to book collections increased. While the libraries of religious institutions and universities – with the exception of the *Ambrosiana* in Milan, the *Angelica* in Rome and the Bodleian in Oxford from the early seventeenth century – long persisted in having fairly restrictive admission policies regarding outside readers,[185] other, newly-established institutional libraries displayed a much more welcoming attitude towards scholars and the general public. Libraries founded in commercial centres such as Florence (c.1440), Venice (c.1470) and Amsterdam (c.1580) opened their holdings to the urban citizenry at large. In Paris between c.1640 and 1720, the *Bibliothèque Mazarine*, the *Bibliothèque de l'abbaye de Saint-Victor*, the *Bibliothèque des avocats* and the *Bibliothèque de la Doctrine chrétienne* and several other libraries likewise began to welcome readers on a regular

184 Petrucci, *Writers and readers*, 174–175, 208–209, 222–224, 226, McDermott, *Social history*, 121–122.

185 Burke, *Social history*, 178, McDermott, *Social history*, 122–123.

basis. From 1692, the royal library in the French capital opened its doors to scholars and the general public. A similar policy was adopted by the *Staatsbibliothek* in Berlin, founded by the Elector of Brandenburg in 1661, and by the library at the Hapsburg Court in Vienna from 1726. Municipal libraries and parish libraries open to locals and outsiders were later founded in many places in Europe.[186] Architects and engineers sometimes built considerable libraries for their own use. Juan de Herrera, architect of King Philip II of Spain, possessed at the time of his death in 1597 a collection of 420 books. In 1576, Pedro Juan de Lastanosa, Philip's master of fortification and supreme mechanical engineer, left a library of 559 books, including many on scientific and technical subjects, like Vitruvius' and Alberti's works on architecture and Frontinus's works on the aqueducts of Rome. This shows, as García Tapia and Carillo Castillo observed, that the 'transmission of technical knowledge in which Lastanosa participated took place to a large extent by means of books and not only by learning on the shopfloor'.[187]

Unlike private collectors in China, moreover, private owners of books in Europe were not at all unwilling to lend out their property. Scholars freely loaned books to friends and even to strangers. The practice of book lending in the eighteenth century even developed into a separate institution in the form of reading clubs and circulating libraries.[188] The holdings of institutional libraries, meanwhile, expanded to include part of the growing body of technical texts and images. In the 1690s, for instance, the university library of Leiden acquired the lion's share of the library left by the polymath Isaac Vossius, which contained, among other things, numerous atlases and books on navigation technology.[189] Institutional libraries thus in a sense turned into 'fixed places' for the collection, storage and dissemination of technical knowledge, although, due to their admission policies and the nature of their holdings, they were in this respect presumably less important than the private collections of technical practitioners themselves (such as that of the Rotterdam millwright, Maarten Nederdijk). This holds true for the libraries of religious organizations as well. Given the prominent role of religious orders in vocational and technical training

---

[186] Petrucci, *Writers and readers*, 225, Jochum, 'Am Ende der Sammlung', 475, Davids, 'Amsterdam as a centre of learning', 317, Burke, *Social history*, 178, McDermott, *Social history*, 124, 126.

[187] Lopez Piñero, *Ciencia*, 245–246, García Tapia and Carrillo Castillo, *Tecnología e imperio*, 76–77.

[188] Tschopp, 'Popularisierung', 475–482, McDermott, *Social history*, 124–126.

[189] Davids, 'In the shadow of Jesuits'.

in several European countries in the seventeenth and eighteenth centuries, one may expect that their schools or houses were well-provided with technological literature. Inventories of the Jesuit library in Antwerp show that this was indeed the case. The Antwerp Jesuits actually possessed an abundance of books and treatises on civil and military architecture, gunnery, surveying, mathematics, optics and mechanics in Latin, Italian, Dutch, French and English.[190]

'Establishing', 'controlling', 'selling' and 'acquiring' technical knowledge in Europe concerned more than knowledge encoded in books, however. Technical knowledge was also carried in the shape of images and objects, both naturalia and artefacts. In the early modern period, these objects were, like books, increasingly being collected, stored and made accessible in fixed places. Botanical gardens, curiosity cabinets, model collections and museums served the same purpose for naturalia and artefacts as libraries did for technical writings. They were stable nodal points in the circulation of technical knowledge. After 1500, all these institutions spread more widely in Europe than in China. Even though equivalents were not entirely absent in the Chinese Empire,[191] they became more common in Europe. In China, collections of technical objects largely remained concentrated in imperial palaces and gardens. In Europe, they spread in many regions and among many social groups.

Botanical gardens, in the sense of institutions holding documented collections of living plants for the purpose of accumulation of knowledge, first emerged in the sixteenth century as a result of systematic efforts by Spanish and Portuguese doctors, apothecaries and missionaries to catalogue the wealth of exotic plants found in the New World and Asia. The massive supply of specimens carried to Lisbon and Seville and the huge amount of information contained in 'natural histories' of overseas worlds composed by scholars such as Nicolás Monardes, García d'Orta and Francisco Hernández gave an impulse to the rise of botanic institutions all over Europe. In Italy, Austria, the Dutch Republic, England, France and other countries, universities, hospitals, apothecaries, urban magistrates or princely authorities started to lay out gardens well-stocked with living specimens of plants from many different parts of the world.[192] The

---

190 Fabri and Lombaerde, 'Architectural treatises', 194–200.

191 Needham, *Science and civilisation*, vol. VI part 1, *Botany*, 331–332, Waley-Cohen, *Sextants of Beijing*, chapters 1 and 3.

192 Cañizares-Esguerra, *Nature*, 27–28, Harris, 'Jesuit scientific activity', Barrera-Osorio, *Experiencing nature*, 111–127.

foundation of these institutes for the collection and public exhibition of plants was not merely curiosity-driven. From the very start, botanical gardens also had obvious utilitarian purposes, both as aids for health care and as sources of income for traders and missionary organizations. Exotic plants were eagerly sought after for pharmaceutical uses and for their potential as ingredients in industrial processes, such as dyeing. From the later eighteenth century, botanical gardens further developed into important centres of knowledge in endeavours to improve agricultural production in Europe and colonial territories overseas.[193] Botanical gardens thus functioned as fixed places at the crossroads of natural and technical knowledge.

Curiosity cabinets, or *Kunst-* and *Wunderkammern*, which arose in Europe during the Renaissance, contained and displayed both natural and artificial 'wonders' of the world. Curiosity cabinets could, according to Mark Meadow, among other things serve as 'sites of cultural and technological production, and active, functional, and practical laboratories for a variety of crafts and disciplines'.[194] Initially, *Kunst- und Wunderkammern* were mainly to be found at princely, noble and ecclesiastical courts. Before long, however, such collections of naturalia and artefacts were also established at educational institutions and in the households of urban elites and professionals.[195]

Like the emergence of botanical gardens, the rise of curiosity cabinets in early modern Europe was made possible by a combination of two factors: the expansion of long-distance commerce and the missionary fervour of the early modern era. One of the richest cabinets in the middle of the sixteenth century could be found at the mansion of the Fuggers, a family of merchant bankers based in Augsburg. Among the exotica in the Fuggers' collection were, for instance, monkeys, parrots, peacocks, leopard skins, orange trees, almond trees, camphor, indigo, pearls, gems, antique coins and scientific instruments.[196] One of the most famous and spectacular *Kunst-* and *Wunderkammern* in the middle of the seventeenth century was the Roman College museum, established in the 1630s by the Jesuit Athanasius Kircher. Kircher's museum was intended as an impressive 'theatre' of the superior role of the Catholic Church in

---

[193] Harris, 'Long-distance corporations', Cook, *Matters of exchange*, 26–28, 111–120, Drayton, *Nature's government*, 50–77.

[194] Meadow, 'Merchants and marvels', 182, 194.

[195] Bergvelt and Kistemaker (eds.), *De wereld binnen handbereik*.

[196] Meadow, 'Merchants and marvels', 183, 191–195.

the collection, storage and dissemination of knowledge about the world, including machines and instruments as examples of human achievements in manipulating nature. Its emergence and growth essentially depended on its connection to the far-flung information and communication network of the Society of Jesus.[197] As a repository of knowledge, the Roman College museum was even richer and more varied than the collection of drawings, engravings and models and other documents accumulated by generations of architects of St Peter's.[198]

Such all-inclusive collections became more unusual, however, as the seventeenth century progressed. Different categories of objects were increasingly housed in separate collections.[199] Naturalia were set apart from artefacts, and within each of these groups, too, differentiation became the rule. Art objects, coins and antiquities were increasingly accommodated in different places from instruments and models of machines. As scaled-down models came to be more frequently used, not only as a means of presentation or as an aid in the design and construction phase of buildings, but also for purposes of instruction, demonstration and experimentation, institutions and individuals began to keep these objects together in separate collections. In the Dutch Republic, for example, repositories of models of machines and other kinds of new inventions were formed by water boards and state agencies such as the States of Holland and the Admiralty of Amsterdam.[200] Soon after its foundation in 1666, the *Académie des Sciences* in France, which among other functions was charged with the task of evaluating the merits of new inventions, also assembled a collection of machines.[201]

While the emphasis in early specialized model collections was on the gathering and documenting of evidence of technical know-how, rather than on disseminating knowledge, a shift of focus became visible from the end of the seventeenth century. Collections of models became increasingly accessible to wider publics. One of the pioneers of this trend was the Amsterdam bookseller, theologian and architectural writer Willem Goeree. In about 1680, Goeree announced that he would open for public view a large repository of 'architectural models and instruments of force

197 Findlen, 'Scientific spectacle'.
198 McPhee, *Bernini*, 36–37.
199 Jochum, 'Am Ende der Sammlung', 273.
200 Davids, *Rise and decline*, 160, 277–278, 432–433, 447, Popplow, *Models of machines, passim.*
201 Hilaire-Pérez, 'Dépôts d'inventions', 35 note 23.

and practice', which included copies of all sorts of buildings and building components, models of locks, sluices, bridges, cranes, drop hammers, windlasses, pumping devices, minting presses and mechanisms of mills.[202] The *depôt d'inventions* maintained by the *Grande Fabrique* of Lyon, which supervised silk production in the city, was also open for public view in the eighteenth century. The *depôt* was deliberately intended as 'un lieu d'apprentissage de gestes et des savoirs'.[203]

Universities and newly-established societies for the encouragement of arts and sciences began to form collections of models for the purposes of instruction and dissemination of knowledge, too. 'Cabinets of physical instruments' in Dutch universities in the eighteenth century comprised numerous models of machines, including small-scale copies of steam engines. The Society for the Encouragement of Arts, Manufacture and Commerce in London opened a repository of machines in 1761, which was publicly accessible. The contents of the collection were described in published catalogues. The *Oeconomische Tak* of the *Hollandsche Maatschappij der Wetenschappen* in Haarlem, the Netherlands, which was partly modelled after the London Society, set up a repository of models of technical devices and implements in 1783. The Museum of *Teyler's Genootschap* in the same city, managed by Martinus van Marum, assembled an exceedingly rich collection of models of machines and instruments in the late eighteenth and early nineteenth centuries, including small-scale copies of capstans, cranes and windmills.[204] The *Conservatoire des Arts et Métiers*, established by the revolutionary government of France in 1794, combined the function of an institute for technical education with that of a public collection of 'machines, modèles, outils, dessins, descriptions et livres dans tous les genres d'arts et métiers'. The newly-formed repository consisted partly of existing collections kept by the *Académie des Sciences* and the *depôt public* of new inventions established in the 1780s, and partly of all sorts of models collected by the French state since the Revolution.[205]

Private individuals, universities, learned societies and state agencies thus took the lead in this new departure in the collection of technical objects from the late seventeenth century. Religious organizations no

---

[202] Goeree, *d'Algemeene bouwkunst*, Preface; Van den Heuvel, 'Willem Goeree', 167–169.

[203] Hilaire-Pérez, 'Dépôts d'inventions', 27.

[204] Davids, *Rise and decline*, 489–495, Hilaire-Pérez, *Invention technique*, 202, Bailey, *Advancement of arts*.

[205] Hilaire-Pérez, 'Dépôts d'inventions', 31, Hilaire-Pérez, *Invention technique*, 268–273, Tresse, 'Conservatoire'.

longer took part; Athanasius Kircher had no successor. In fact, the Roman College museum did not long survive its founder's death in 1680.

Conclusion

Throughout the period between c.700 and 1800, movements of people in China and Europe definitely remained the most common medium for the circulation of technical knowledge. However, this chapter has also demonstrated that there were differences in patterns of circulation between China and Europe and within Europe, and that these differences tended to vary in the course of time. Some of these differences concerned the scope and intensity of movements of people, others related to the relative importance of non-human carriers of knowledge – texts, visual representations or artefacts – for the way in which the collection and storage of technical knowledge took place and for the manner in which technical information was made accessible to different groups of people. And the roots of these differences, especially those between China and Europe, can partly be found in variations in religious contexts.

In the Chinese Empire, movements of people relevant to the circulation of technical knowledge were dependent on state regulation for a longer time and to a greater extent than in Europe. Arrangements like the examination system for state officials and the corveé system for artisans, which both involved the regular circulation of people throughout the empire, had no equivalent in the world of Latin Christendom. However, the tramping of journeymen, which occurred independent of state regulation, became a widespread practice in many parts of Europe from the Late Middle Ages, whereas it never came into fashion in Ming or Qing China. After 1000, it also became more common for monks and pilgrims in Europe than in China to travel beyond the immediate surroundings of a village, town or district.

In China, however, technical information circulated much earlier in script and print, and in combinations of texts and images, than in Europe. This mode of circulation was already significantly advanced in China by the beginning of the fourteenth century. In Europe, by contrast, manuscripts with technical information in verbal and pictorial form did not begin to spread until the fifteenth century, and diffusion in print started about a hundred years later.

The differences between China and Europe increased after 1500. While China saw the state-regulated system for artisans decline (though the examination system for state officials remained buoyant), tramping by journeymen continued on a large scale in Central Europe, France and England throughout the early modern period. Migration of skilled exiles and refugees contributed to the circulation of technical knowledge across the Continent. Long-distance travel by Europeans was more common than travelling by Chinese in the Late Ming and early Qing. And from the late seventeenth century, a new genre of travelling in Europe arose, specifically aimed at promoting the circulation of technical knowledge: technological journeys.

The most striking divergence, however, took place with regard to non-human carriers of knowledge and in the collection, storage and accessibility of technical data. In technical writing about mechanics and machine-building, by the early seventeenth century, Europe had become an example for China instead of the other way around. Technical illustrations in Europe, in contrast with China, steadily became more realist, detailed and precise. European 'machine-books' began to record both technology-in-use and projected images of technology that might be developed in future. Encyclopaedias and technical journals, which started to appear in Europe in increasing numbers from the eighteenth century onwards, contained more up-to-date information on technology than similar publications in China and also reached a much wider audience. In Europe, artefacts, especially models, became an ever more important medium for the circulation of knowledge, and not just in architecture and the building industry. Moreover, the network of 'fixed places' where technical information was assembled and preserved assumed a different structure in Europe from that in China. Libraries in early modern Europe became more prolific and more easily accessible to various groups of people than in Qing China. In the early modern period, objects were, like books, increasingly being collected, stored and made accessible in fixed places. Botanical gardens, curiosity cabinets, model collections and museums served the same purpose for naturalia and artefacts as libraries did for technical writings. After 1500, all these institutions became more widespread in Europe than in the Chinese Empire.

This growing divergence in the circulation of technical knowledge between China and Europe can in part be related to variations in religious contexts; in part, but not entirely. The circulation of technical knowledge by movements of people and non-human carriers of knowledge in China and Europe was also driven by the actions of state agencies, scholars, commercial publishers, merchants, engineers and craftsmen, as we have seen, and so was the collection and storage of information in various kinds of 'fixed places'. An important reason why Europe began to diverge from

China in this respect may very well have been that in Europe, the latter groups had more opportunity to act on their own, irrespective of guidelines from governments, than in the Chinese Empire, where the 'shadow of hierarchy'[206] from state agencies continued to loom much larger.

Yet, religious institutions, rituals and practices made a vital contribution to the divergent evolution of the circulation of knowledge in Europe, both before and after 1500. In the High and Late Middle Ages, pilgrimage and regular communication between formally 'networked' monastic houses in Europe was, in all probability, already facilitating long-distance flows of technical knowledge. Such conduits for the circulation of knowledge were much less developed in China. From the sixteenth century onwards, the Protestant and Catholic Reformations did not, on balance, tend to obstruct the circulation of technical knowledge. Although at times, these religious transformations led to massive waves of 'confessional migration' of skilled people and in many places had an influence on admission policies for master-craftsmen, the actual impact of these changes on regional or local technological development was not very marked, either in a positive or a negative sense. Movements of journeymen, meanwhile, were hardly affected by the newly-emerged religious boundaries at all. The 'Little Divergence' in economic performance between areas in north-western Europe and other regions of the Continent, which often happened to be Catholic, can no more be ascribed to some retarding impact from the Tridentine Church than in the case of human capital formation. In fact, the Catholic Reformation even boosted the circulation of technical knowledge in important ways. After 1500, missionary fervour generated a huge increase in long-distance travelling by Europeans, which in turn led to a remarkable expansion and acceleration in the circulation of technical knowledge and skills.

As with regard to human capital formation, the difference in the impact of religious contexts between China and Europe may essentially be explained by the varying relations between religious institutions, governments, markets and demand 'from below'. In the interplay between these variables, religious institutions in Europe could act more independently than in China. In many times and places, networks of monastic houses or many-branched missionary organizations, such as the Jesuits, drew benefit from support by princes or states, but, up to the end of the eighteenth

206 As political scientists would call it nowadays, cf. Scharpf, 'Games real actors could play'.

century, they could also very well go their own way – and survive and thrive. And what they did more often than not aided the smooth circulation of technical knowledge.

Figure 3. Frontispiece of Kaspar Schott, *Technica curiosa, sive mirabilia artis*, published in Nuremberg in 1664. The picture shows mathematical instruments and artefacts (including a copy of the Magdeburg hemispheres, top left) discussed in this book by Schott, a Jesuit scholar based in Würzburg, Germany. This particular copy formed part of the library of Isaac Vossius, bought by the University of Leiden in 1690. Courtesy Leiden University Library, The Netherlands, shelf mark 540 F 4.

CHAPTER FOUR

RELIGION AND TECHNICAL INNOVATION IN CHINA AND EUROPE

Human capital and the circulation of knowledge and skills lie at the start and the finish of the 'loop' of technological change. Human capital is both an ingredient and a product of technological change – and the same applies to the circulation of knowledge and skills. Having discussed differences in human capital formation and the circulation of technical knowledge between China and Europe and within Europe in the previous chapters, I will now concentrate on the middle section of the loop: technical innovation. Technical change, after all, implies the appearance of new things, whether fast and radically or slowly and incrementally.

Technical innovation is itself a complex affair. It involves an intricate combination of incentives, settings, resources, mental operations and design practices. To understand the creation of technical innovations we thus have to consider a variety of relevant elements: the expected uses and rewards for developing new things, the settings and resources from which new things can be fashioned, the mental steps by which ideas for innovations can be turned into reality and the practical ways in which these can be shaped into techniques or artefacts that actually work. Although the present state of inquiry on technological change in China and Europe does not yet allow for as detailed an examination of each of these elements as one might wish, in this chapter I will nevertheless seek to demonstrate that differences in the process of technical innovation in China and Europe *did* emerge, that these differences, especially after 1500, were more evident *between* China and Europe than *within* Europe, and that they have been affected by variations in religious context in significant ways.

Religious contexts are conceived here in the full sense discussed in the introduction of this book. Religion is taken as a multi-faceted phenomenon, which finds expression in people's beliefs, practices and experiences, in institutions and media as well as in rituals, images and doctrines. These expressions are all based on the assumption – in Steve Bruce's words – 'of the existence of supernatural entities with powers of action, or impersonal powers or processes possessed of moral purposes'.[1] The key question addressed in this chapter is to what extent variations in religious contexts, conceived in this sense, may help to explain differences in technical innovation in China and Europe.

The introduction of this book explained that differences in technological evolution between China and Europe cannot be reduced to Chinese 'superiority' versus European 'backwardness' before 1500, or to Chinese 'standstill' versus continued European 'advance' in early modern times. Between c.700 and 1800, both China and Europe went through long periods of incremental technological change, interspersed with outbursts of accelerated creativity. The fact that a revolutionary technological breakthrough was eventually achieved in Europe and not in China should not induce us to exaggerate differences in the *level* or *rate* of technological change in Europe and China before the end of the eighteenth century. Neither ought we to assume, on the basis of the fact that this breakthrough was first realized in Britain, that vast differences in the *level* or *rate* of technological change before that time existed between regions in Europe. Variations in technology within Europe before the mid-eighteenth century were never that large, and the gap opened by the acceleration in Britain could quite rapidly be bridged by other countries in Europe in the first decades of the next century. Yet, even if the level and rate of technological change in China and Europe before the end of the eighteenth century did not vary markedly, there were differences in the *nature* of technological change. Variations in technological evolution between China and Europe concerned the degree of development, or reversals in development, of particular techniques or artefacts, as well as the emphasis by sector of technological advance. These variations reflected a growing divergence in the creation of technical innovations.

This chapter argues that this increasing divergence in technical innovation between China and Europe was related in important ways to variations in religious contexts. Religious differences between China and Europe, as the first section of the chapter shows, mattered in incentives and settings as well as in resources, mental operations and design practices. The next section of the chapter aims to demonstrate that variations *within* Europe can more readily be explained by political, social or economic changes than by religious factors. While fault lines between Protestant and Catholic regions were only of minor importance, variations in

[1] Bruce, *Religion*, 7.

religious contexts can go some way to explain differences in technological development too. Variations in religious contexts were mainly important in so far as they affected differences in development between countries where Protestantism was the dominant confession.

Religion, Technical Innovation and the Great Divergence

Religious Institutions as Innovators

One of the most memorable illustrations of the interconnections between religious contexts and technical innovation in European history is David Landes' account of the early development of the mechanical clock. In his *Revolution in time*, Landes traced the interest in devices for time measurement back to the concern for punctuality in medieval monastic houses. The daily rituals of monastic life in Latin Christendom, especially under the reformed regime from the eleventh century, exacted a strict temporal discipline. Mechanical clocks were an effective means to serve this need. Although the first reports of mechanical clocks in Europe do not appear until the fourteenth century, Landes confidently claimed that before that time, it was the monastic clergy who 'provided the primary market for timekeepers and the principal stimulus to technical advances in this domain'.[2]

An equivalent to this story in China might be the tale of the emergence of woodblock printing. The oldest surviving examples of woodblock printing in China, Japan and Korea are Buddhist charms from the eighth century. The oldest extant example of a book printed with woodblocks is a copy of a Buddhist text from 868, the Diamond Sutra, found among a hoard of documents at the Dunhuang monastic complex in Northwest China.[3] In its early stages, woodblock printing was apparently closely connected with Buddhism. The explanation, as John Kieschnick and Michael Mitterauer suggest, may lie in a combination of three factors. Buddhism, as a religion of scripture, always set great store on the reproduction and diffusion of sacred images and texts. The copying and multiplying of religious images and texts was regarded as a highly praiseworthy activity. Buddhism was also a missionary religion. Buddhists were eager to spread their message to new audiences, both in the original home of Buddhism,

---

[2] Landes, *Revolution in time*, 53, 58–70, cf. Bruton, *Clocks*, 17.
[3] McDermott, *Social history*, 11.

India, and beyond. The region where Buddhism made its greatest gains between the sixth and early ninth centuries, Tang China, moreover offered excellent conditions for the large-scale reproduction of images and texts, because of the easy availability of a material to work upon, namely paper.[4] Religious stimuli combined with suitable material conditions thus formed a favourable environment for the development of a new technique of reproduction, woodblock printing.

Religious institutions also formed a friendly environment for innovation in the European building industry. Nowhere in Europe in the High and Late Middle Ages were places to be found that continuously offered such a favourable ambience for experiment as the construction sites of cathedrals. Gothic cathedrals, David Turnbull has observed, can be 'conceived as sites of experimental practice: literally as "laboratories"'. One of the ways in which cathedrals functioned as laboratories was that during the process of construction, 'close observation of the drying mortar enabled the builders to detect areas of stress in the fabric and to take appropriate remedial measures through the placement of buttresses, pinnacles, or reinforcement'. 'The use of templates', moreover, 'along with constructional geometry and a relatively small range of simple tools – compasses, straightedge, and string – [enabled] in an experimental context, the construction of extremely high, radically innovative buildings'.[5]

In China, religious institutions similarly had a stimulating influence on innovation in the building industry. In all probability, the temple building boom, which lasted from the Tang to the Yuan dynasties, made a significant contribution to the perfection of what Lothar Ledderose labelled the 'modular' system in construction. A typical feature of Chinese production systems, according to Ledderose, was the 'assembl[ing] of objects from standardized parts', which were 'prefabricated in great quantity and could be put together quickly in different combinations, creating an extensive variety of units from a limited repertoire of components'. In the building industry, the 'parts of the full-fledged [modular] system were first joined in organic unity' during the Tang dynasty.[6] As many of the larger buildings erected in this period were temples, it is very plausible that the sustained construction drive of these religious edifices helped in bringing this design practice to the highest state of development.

---

[4] Kieschnick, *Impact of Buddhism*, 167, 172, Mitterauer, 'Religion und Massenkommunikation', 246, 256–258, cf. also Elvin, *Pattern*, 184.

[5] Turnbull, 'Ad-hoc collective works', 315–316, 321–322.

[6] Ledderose, *Ten thousand things*, 1, 117, 120.

Both in Europe and in China, religious institutions were initially in the forefront of innovation in hydraulic technology. Along with palaces, monasteries in Europe were the first places to build complex systems for water supply in the High Middle Ages. In the eleventh and twelfth centuries, fully-fledged water systems with conduits, cisterns, pipes and fountains in *lavatoria* were installed in numerous monastic houses on the Continent and in the British Isles.[7] 'Religious houses seem to have been more willing than other potential sponsors to construct complex water systems', Roberta Magnusson has remarked. One of the reasons for this innovative attitude may, again, have lain in the ritualized nature of monastic life. Magnusson suggested that 'the requirements of... a closely, regulated, synchronized communal life', which led to 'exceptionally high peak demands' for water, helped to 'create a pre-existing need that favored' the adoption of such complex systems.[8]

Temples in China, especially on mountaintops, appear to have represented masterpieces of hydraulic engineering, too. Jesuit missionaries in the early eighteenth century were very impressed; monks had made temples into sites of 'eternal spring', father Laureati reported to Europe in 1714. 'Ils font couler les eaux du haut des montagnes par plusieurs canaux, et ils les distribuent aux environs et dans l'interieur de la pagode, où il y a des bassins et des fontaines pour les recevoir'.[9] It is not unlikely that Buddhist monks were also innovators in bridge-building. From the Song onwards, many records named specific monks, such as Huaibing and Xuanyun, as being particularly ingenious in restoring old bridges or designing new ones.[10]

Religious and Secular Settings of Innovation

After about 1500, however, the impact of religious contexts on the emergence of new inventions in China and Europe began to diverge in significant respects. In the sixteenth and seventeenth centuries, religious institutions in Europe contributed much more to technical innovation than temples or monasteries in contemporary China. This relative difference in weight was closely connected with the unprecedented amount and concentration of church support for the arts and the building industry in the heart

7 Magnusson, *Water technology*, 5–6, 13–15.
8 Magnusson, *Water technology*, 21–22.
9 Vissière (ed.), *Lettres*, 207.
10 Kiesnick, *Impact of Buddhism*, 204–205.

of Latin Christendom, papal Rome. Once the papacy, after a long period of exile and schisms, had finally and permanently re-established its seat in Rome in the early fifteenth century, ecclesiastical authorities embarked on a huge, long-term project of extending, reconstructing and beautifying the city. The grand vision guiding popes, cardinals and bureaucrats in their sustained endeavours was the ideal of turning the city on the Tiber into an earthly image of the heavenly Jerusalem. Restoring the grandeur of the former capital of the empire and embellishing it with buildings 'seemingly made by the hand of God', the papacy would transform Rome into a dazzling symbol of its claim to supreme spiritual and temporal power.[11]

This vast, religion-inspired urbanist programme ranged all the way from improvement of public utilities and restructuring of public space to the erection of imposing monuments and the construction of glorious residences and magnificent religious buildings. Legions of craftsmen, artists, architects and engineers, supervised by a steadily growing papal bureaucracy, provided the expanding city with a myriad of new streets, squares, fountains, fortifications, gates, bridges and river embankments.[12] Public spaces were adorned with statues and monuments reflecting the bonds between papal Rome, Antiquity and Eternity. Erecting these structures and artefacts often involved extraordinary and innovative feats of engineering. Faced with the challenge of raising columns, obelisks and towers, in the fifteenth century, the Siennese engineer Francesco di Giorgio thought up all sorts of ingenuous solutions in his drawing book, which were copied time and again in manuscripts for nearly 150 years.[13] The removal of the Vatican obelisk to a position in the square facing the new St Peter's basilica in 1585–1586, under the supervision of architect-engineer Domenico Fontana and with the help of hundreds of men and horses powering dozens of winches and pulleys, was a first-rate technological achievement, publicized and studied all over Europe during the late sixteenth and seventeenth centuries.[14] Moreover, popes, cardinals and religious orders commissioned the construction of dozens of churches, convents, monasteries and *palazzi*. Between 1527 and 1590 alone, Rome saw the building of more than 50 new churches.[15] The greatest enterprise of all was, of course, the construction of the new St Peter's, with its adjacent square

---

[11] Partridge, *Renaissance Rome*, 20–21.
[12] Partridge, *Renaissance Rome*, 20–26, Long, 'Hydraulic engineering'.
[13] Galuzzi, *Renaissance engineers*, 164–169.
[14] Sawday, *Engines of the imagination*, 59–68.
[15] Partridge, *Renaissance Rome*, 56.

and embracing colonnade, which started in 1506 and was more or less completed in the 1670s.

Among the craftsmen, artists and technical experts who carried this massive programme of urban transformation into effect were some of the most creative minds of the time, including *virtuosi* like Raphael and Michelangelo. Their path-breaking achievements did not only spring from their own ingenuity, but also from their access to resources, ideas and practices accumulated in Classical Antiquity or stored in contemporary repositories of knowledge and skills. In the early sixteenth century, Donato Bramante, the first architect of the new St Peter's, could benefit from the recent 'recovery' of the ancient Roman technique of 'poured concrete, formed and cast on site'.[16] One of his seventeenth-century successors, Gian Lorenzo Bernini, who designed the square, the colonnade and much of the interior of the basilica, possessed jointly with his brother Luigi, an accomplished engineer and mathematician, a well-stocked library of books and manuscripts by classical and contemporary authors on architecture, machines, mechanics, optics, geometry, arithmetic and painting. Among the items in their collection were titles by Vitruvius, Serlio, Palladio, Vignola, Sabbattini, Euclid, Tartaglia and Galileo.[17]

Little of this programme in Rome would have been realized without the sustained support of ecclesiastical authorities, however. And it was the system of beliefs, doctrines and rituals of the reformed Catholic Church that left its stamp on many of the new creations – most of all, of course, on the design of religious buildings. When designing new churches, for example, architects and engineers had to carefully calculate light and sound effects and spatial arrangements in such a way that these buildings would be optimally suited to the holding of frequent masses and preaching to large congregations. The *Gesù*, the principal church of the Jesuits in Rome, completed in 1584, became a classical example of how successfully this aim could be realized.[18] But the influence of innovations in Rome reached far beyond the city – and not just because the design of buildings like the *Gesù* was widely imitated in Catholic churches elsewhere in Europe, the Americas and Asia. Rome at the dawn of the early modern age constantly attracted a large number of pilgrims: an estimated 30,000 in a normal year, and some 100,000 in a Jubilee Year, which from 1450

16 Partridge, *Renaissance Rome*, 51.
17 McPhee, *Bernini*, 41–42.
18 Partridge, *Renaissance Rome*, 57–58.

took place once every 25 years.[19] Many Catholic Europeans would thus have seen this 'model city' with their own eyes. And non-Catholics sometimes had a keen interest, too. English *cognoscenti* like mathematician John Wilkins, architect Inigo Jones and author John Evelyn (who travelled in Italy in the 1640s) were all highly impressed by Fontana's account of the transport of the Vatican obelisk.[20] The architect of London's St Paul, Christopher Wren, modified his design of the dome of the new cathedral after hearing about the unforeseen problems of cracking in the dome of St Peter's that had come to light by the 1690s.[21]

Although for a long time, religious organizations in Europe thus continued to have a greater impact on the creation of new technical inventions than their counterparts in China, they nevertheless gradually lost ground to secular actors and institutions. After the fourteenth century, the centre of innovation in clock-making no longer lay in monastic houses but in urban communities. Towns in Central Europe, Italy, France and the British Isles vied with each other to have the most elaborate and intricate public clockworks built within their walls. 'A show clock was a matter of prestige, an edifying spectacle for residents and visitors alike', according to David Landes.[22] Such clockworks were often installed, by order of the city authorities, in the outer wall of a secular public building, such as a gate, a belfry or a town hall, as was the case in Bern, Rouen or Olomouc; or inside the largest, most spacious building of the city, a cathedral, as in Strasbourg. Monastic houses did not remain the sole leaders in hydraulic technology either. During the rapid urban expansion of the twelfth and thirteenth centuries, many cities in Italy, France, the British Isles and other regions began to build complex water systems, too, sometimes in direct collaboration with religious institutions. The construction of aqueducts, pipes, fountains and so forth was both a response to the growing need for fresh water and a manifestation of increasing inter-urban rivalry.[23] Cities did not only attempt to trump each other with awe-inspiring clockworks, but also with stunning fountains.

A more complex pattern manifested itself in organ building. By 1450, permanent pipe organs – as distinct from conventional, portable organs – had become in 'most representative types of church' (including cathedrals

19 Partridge, *Renaissance Rome*, 16.
20 Sawday, *Engines of the imagination*, 67–68.
21 Mark, 'Technological innovation', 21–23.
22 Landes, *Revolution in time*, 79–80, 83, 191–200.
23 Magnusson, *Water technology*, 22–33.

and monastery churches) in most parts of Latin Christendom a conventional instrument to accompany ceremonies of worship.[24] Organ building, like clock-making or hydraulic engineering, developed into a highly specialized craft, which combined the command of a wide spectrum of skills – ranging from drawing and tuning to joining wood, casting metal and handling leather – with the ability to satisfy a large variety of specifications for individual instruments. The organ builder became a kind of accomplished machine maker. When Europe fragmented religiously in the sixteenth and seventeenth centuries, churches remained the most common location for these mighty devices, but the diversity in uses and design increased. Permanent organs, even those installed in churches, became less exclusively linked to religious uses. Some cities or countries, after turning Lutheran, Calvinist or Puritan, banned the use of church organs altogether.[25] In other places, both in Protestant territories such as the Dutch Republic, and in regions that remained faithful to the Catholic Church, such as Northern Italy, organs were frequently used not only for liturgical purposes, but also for the performance of popular recitals, oratorios and operas. For urban magistrates in the Netherlands, building large, powerful organs in publicly-owned churches served a similar function to the construction of public clockworks in medieval towns: showing off the status and prosperity of their own city vis-à-vis other towns. And these secular authorities did so *against* the initial opposition from Reformed ministers.[26] In yet other regions in Europe, such as Protestant and Catholic territories in the Holy Roman Empire, the use of organs commonly remained restricted to religious ceremonies. Rivalries could occur, too, but these were often of an ecclesiastical nature. In Catholic territories, religious orders such as Jesuits, Augustinians, Cistercians, Benedictines and Dominicans each cultivated their own traditions of organ building in their churches and abbeys.[27]

Increasingly, however, technical innovation in Europe also occurred in settings that were not, or barely, related to religious organizations. The expansion of the market economy, the rise of urban noble courts and the growth of patronage by secular governments opened a new space for a

---

[24] Williams, *Organ in western culture*, 361, Van Eck, 'Geschiedenis', 119.

[25] Williams, *European organ*, 67, Norman, *Organ today*, 8, Van Eck, 'Geschiedenis', 130–131.

[26] Williams, *European organ*, 27–30, 36–37, 40, 206–206, 217–218, Van Eck, 'Geschiedenis', 163–169.

[27] Williams, *European organ*, 58.

creative class of technical practitioners and artisans.[28] While Michelangelo, Bramante and the Berninis spent much of their creative lives working for cardinals and popes, there were countless other social or institutional environments where highly-skilled and gifted people found ample opportunity to deploy their talents. Leonardo da Vinci realized many of his greatest feats as an artist, architect and engineer while enjoying the patronage of the Duke of Milan, the King of France and magistrates of Florence and Venice. Architects and engineers from the Late Middle Ages became professionals in their own right and reached a status far above the rank and file of mere masons, carpenters and blacksmiths. Painters or sculptors began to claim a special place for themselves, too, contending that they possessed a singular ability to know nature through the practice of their art. 'Machine-books' and other texts and sketches produced by engineers and artisans gave an eloquent verbal and visual testimony of their skills and imaginative powers.[29] 'It was the recognized existence of expert makers', in David McGee's view, that formed 'the very ground of the individual creativity we see in early machine drawings'.[30]

The expertise of these practitioners and artisans in designing new things – whether a building, a fountain, a new kind of painting or a new mechanical device for civil or military use – became highly sought-after among princes, noblemen, urban governments and church authorities alike. Members of this creative class could be found moving from place to place in Europe – and not just the 'top players' like Da Vinci, Francesco di Giorgio or Albrecht Dürer. Among all of the engineers active in Spain in the sixteenth century, for example, no more than 75% came from Spain itself; some 15% hailed from Italy and another 10% from Germany, Flanders, France and England.[31] The new fortifications built around cities in the Low Countries between c.1530 and 1570 were largely designed by engineers and architects from Italy, paid by the Hapsburg government or municipal authorities. During the first phase of the Dutch Revolt, the army of the States General enlisted the service of a number of brilliant Italian engineers, such as Nicolas Romero and Frederigo Gianibelli.[32] Dutch

---

[28] Cf. Smith, *Body of the artisan*, 25, 238.

[29] Turnbull, 'Ad-hoc ', 330–331, Mark, 'Technological innovation', 18–19, 25, Popplow, 'Why draw machines', 28, Hall, '"Der Meister sol auch kennen schreiben und lesen"', Dohrn-van Rossum, 'Migration-Innovation-Städtenetze', Lindgren, 'Ordnungsprinzipien', 205–214, Long, *Openness, secrecy, authorship*, chapter 4, Lohrmann, 'Das Maschinenbuch'.

[30] McGee, 'Origins', 84.

[31] García Tapia, *Ingeniera y arquitectura*, 41.

[32] Davids, *Rise and decline*, 213.

experts in hydraulic and military engineering, in their turn, were in great demand with governments and private entrepreneurs all over Europe throughout the seventeenth and eighteenth centuries.[33]

Technical innovation in China, to a greater extent than in Europe, took place *outside* the ambience of religious institutions. But market demand for architect-engineers or artisans hardly developed. Engineers were present in force, but they were mainly active as government employees. Take, for example, a central area of engineering activity in China: river control. From an early date, river control in China formed part of the concerns of the imperial state. The technical management of river control was carried out by an ever-expanding hydraulic bureaucracy, assisted by provincial officials for the organization and supervision of maintenance and repair jobs at a local level. A central agency to coordinate efforts to control the Yellow River, the Office of Rivers and Canals, was established in the middle of the eleventh century. As Christian Lamouroux has observed,

> The centralization of the management of the resources destined for the handling of the river was shown indispensable once hydraulic operations, whose costs could no longer be borne at the local level, appeared as a new charge in the budgets controlled by the central administration.[34]

In the following centuries, the imperial government considerably stepped up the input of resources for the maintenance of the control system of the Yellow River, especially since from the 1410s onwards, the management of the river was closely connected to the upkeep of the Grand Canal. The Grand Canal, first constructed under the Sui and Tang dynasties, and redesigned under the Yuan and Ming, served as the main artery for the supply of foodstuffs and other goods to the capital and army in the North. The north-south route of the Canal 'included a portion of the lower course of the Yellow River, a fact that complicated both canal transport and river management'.[35] Because of its vital importance for the preservation of the state itself, the maintenance of the Yellow River and Grand Canal system became one of the principal preoccupations of the imperial bureaucracy. The efforts to keep the control system intact, despite the growing problems caused by the steady accumulation of silt left by the river, were not abandoned until the middle of the nineteenth century, when the Qing

---

33 Davids, *Rise and decline*, 286–320.

34 Lamouroux, 'From the Yellow River to the Huai', 559–560.

35 Leonard, *Controlling from afar*, 23–36, Dodgen, *Controlling the Dragon*, 15.

state 'could no longer afford to keep [it] operating'.[36] In the Qing period, the top positions in this hydraulic bureaucracy were increasingly filled by people who had risen through the ranks and had thus acquired a high degree of specialization in river hydraulics.[37]

The expansion of this imperial hydraulic bureaucracy was clearly related to the fact that the problem of controlling the Yellow River had become inextricably connected with that of managing the chief supply route to the capital of the Empire, the Grand Canal. In the basin of the middle Yangzi, where such a direct connection did not exist,[38] the role of this bureaucracy was much smaller. Although government subsidies were a significant stimulant to the extension and improvement of dyke construction in the plains of Hubei, Hunan and Jiangxi during the Song, the early Ming and the reign of Emperor Kangxi (1661–1722), state control of river management in this area did not reach the same level of intensity as in the case of the Yellow River basin. Local communities and private individuals always played a significant part in dyke building and dyke repair, legally or illegally.[39] Supervision by higher officials was strengthened after a disastrous flood in 1788, but these measures did not result in the permanent growth of bureaucratic power in hydraulic affairs in this region.[40] But even in this field, a market for a creative class of engineers did not emerge.

Incentives for Innovation

It was not only the settings for the creation of technical inventions in Europe that grew more diverse after 1500; the diversity of the incentives to do so increased, too. Incentives for technical innovation in Europe were also introduced outside the existing ambience of religious institutions or outside the context of emulation between urban and religious communities. A typical feature of these new incentives was that they consisted of a mixture of individualized rewards and provisions to benefit the community at large. According to North and Thomas' theory of economic growth, innovation will be encouraged by organizing the institutional

---

[36] Dodgen, *Controlling the Dragon*, 159.

[37] Dodgen, *Controlling the Dragon*, 22–24.

[38] There was an indirect connection in the sense that from the early sixteenth century, this region became an important exporter of rice, which found its way down the Yangzi and thence could reach the Grand Canal.

[39] Will, 'Un cycle hydraulique', 268–281, Perdue, 'Water control', 752–759.

[40] Perdue, 'Water control', 759–761.

environment in such a way that the private rate of return on inventive activities approaches the social rate of return. In their view, individuals or groups will only be willing to undertake substantial expenses for devising new things if the property rights to their intellectual products are reasonably ensured. If not, they will not be prepared to bear the costs of sustained research efforts. On the other hand – thus their argument runs – institutional arrangements have to be balanced in such a way that society, too, will reap the benefits of the inventions of individual members. If not, growth will not be forthcoming.[41] Although North and Thomas' assumption about the need for protection of the private rate of return as a precondition for inventive activities does not universally hold – witness the case of 'collective invention'[42] and the frequency of technical innovations in the context of religious organizations discussed above – it is nonetheless true that individualized, material recompense became an increasingly important component of incentives for innovation in Europe.

One of these new forms of incentive was the patent. Granting patents for technical innovations is a way to open up new knowledge to wider groups of people, on condition that the inventor will have sufficient opportunity to receive adequate rewards for his labour. Patent systems did not exist in Imperial China,[43] but they *did* arise in late medieval Europe. Pamela Long has argued that awarding patents for inventions was an expression of increasingly proprietary attitudes towards craft knowledge, which was related to the rise of cities in the High Middle Ages. Patents for mechanical devices or craft techniques were granted in Italian cities from the thirteenth century onwards. The criterion for granting a patent at that time was not the originality of an invention, but its novelty in a particular locality. This was also the principal requirement laid down in the first law guaranteeing an inventor's rights to the ownership and commercial exploitation of a newly introduced process or contrivance, issued by the Senate of the Republic of Venice in 1474.[44] Other European states followed the example of the Italian city-republics. In the Kingdom of Castile, patents first appeared at the end of the fifteenth century. They became more frequent after the task of examining inventions and granting patents in

41 North and Thomas, *Rise*, 2–7, 154–155.

42 Allen, 'Collective invention', Davids, *Rise and decline*, 389–390, 395–398.

43 Liu, 'Reflections'.

44 Long, *Openness*, 89, 93–96, Molà, *Silk industry*, 186–189, Mandich, 'Le privative industriali Veneziane', 537–547, Belfanti, 'Guilds', 577–580, Biagioli, 'From print to patents', 141–151.

the Kingdom was entrusted to the Council of Castile. The first patent awarded by this state agency dates from 1522. The Council of the Indies soon started to grant patents, too. The first patents in the Hapsburg Netherlands and in England were granted in the 1550s and in Scotland in the 1560s.[45] The French state began to award *privilèges exclusives* to inventors in the mid-sixteenth century and after c.1670, entrusted the task of assessing new proposals to the *Académie des Sciences*.[46]

Other forms of individualized, material stimuli for creating new knowledge that spread in Europe in the early modern era were rewards, prizes or premiums and sponsorship by public or semi-public institutions by means of commissions, contracts and other sorts of privileges. Rewards, prizes or premiums could, to some extent, fulfil the same function as patents. They offered a material benefit for an inventor without giving him the right to keep his invention secret forever. The difference was, of course, that this benefit could be cashed more directly than by means of a patent right, that there were no temporary restrictions on the use of the invention by people other than the inventor (or his representatives), and that a system of rewards, prizes or premiums made it possible to encourage inventive efforts in a more direct, specific way than a patent system, which by its very nature only provides a passive, generic sort of support.

A good example of this kind of incentive is that of longitude prizes. Spain, again, led the way. Shortly after his accession as King of Spain in 1598, Philip III offered a reward for anyone who could come up with a practical and reliable solution for the problem of finding longitude at sea, consisting of a sum of 6,000 ducats plus a life annuity of 2,000 ducats and a further 1,000 ducats for expenses. The *Casa de la Contratación* in Seville was entrusted with the task of judging the merits of the proposals that were submitted. Hardly had the Spanish Crown offered its longitude prize than its opponent in the Low Countries, the Dutch Republic, made a similar move. On 1 April 1600, the States General of the United Provinces offered a sum of 5,000 guilders plus a life annuity of 1,000 pounds for anyone who would present an adequate solution for the problem of finding what the Dutch called 'the East and West'. A year later, the States of Holland likewise offered a longitude prize. In the Dutch case, the merits of the proposals were usually assessed by ad-hoc committees of experts. In Great

[45] Mateos Royo, 'State policy', 149–150, García Tapia, *Tecnica y poder*, 195–201, Davids, *Rise and decline*, 401–402, MacLeod, *Inventing the Industrial Revolution*, 11.

[46] Hilaire-Pérez, *L'invention technique*, 42, 47–50.

Britain, a longitude prize was established by Act of Parliament in 1714 and administered by a newly-created Board of Longitude. The *Prix Rouillé* for new inventions in commerce and navigation in France, instituted in 1715 and administered by the *Académie des Sciences*, fulfilled a similar function as the prizes in Spain, the Dutch Republic and Britain.[47] Although practical, accurate and reliable solutions to the problem – relying partly on the principles of the mechanical clock invented way back in the Middle Ages – were not found until about 1770, longitude prizes generated a wave of inventive activities in Europe throughout the seventeenth and eighteenth centuries.

Awards for technical innovations were granted by many sorts of institutions: at first mostly by state agencies, urban governments and other public institutions, such as the *Caisse du droit des étoffes étrangères* in Lyon, later also by private or semi-public associations such as the Society of Arts in Britain, the *Oeconomische Tak* in the Netherlands, the *Société d'Émulation de l'abbé Baudeau* in France or agricultural, scientific, 'improvement' or 'patriotic' societies, which mushroomed in many countries in Europe in the later eighteenth century.[48] Commissions, contracts and other sorts of privileges awarded by government agencies or by chartered companies such as the Dutch East-India Company could also provide individual inventors with tangible, material recompense for their efforts. Inventors could also benefit from remuneration from a combination of sources. In the 1670s, for example, the Dutch painter Jan van der Heyden received several patents from the States General and Provincial States for an improved type of fire-fighting device consisting of a suction pump, a pressure pump and a leather (later canvas) hose. Van der Heyden also obtained a patent on a new kind of water-raising mill from the States of Holland and the States General. Moreover, during the late 1660s and 1670s, he devised a number of inventions for which he never requested patents at all: a mud-mill, a stove and a new system for the lighting of cities. Apart from income from his patents, by selling copies of his fire engines to a variety of clients ranging from cities in the Netherlands and Germany to the Dutch East-India Company and the Elector of the Palatinate, Van der Heyden found a solid and enduring financial basis for his activities by entering into a long-term relationship with the government of Amsterdam. Following the decision of the

[47] Jonkers, 'Rewards and prizes', 434, Marguet, *Histoire*, 48–49.

[48] Hilaire-Perez, *L'invention technique*, 71–74, 190–220, Davids, *Rise and decline*, 123, 410–413, 456, 481–482, Endres, 'Handwerk- Berufsbildung', 403, Hubrig, *Patriottische Gesellschaften*, 48–51.

burgomasters in 1669 to adopt his newly proposed system for city lighting, Van der Heyden was appointed supervisor-general of city lighting which, from 1670 until his death in 1712, yielded a steady income of 2000 guilders a year.[49]

Was there anything like this in China? Invention was appreciated, it is true, even to the extent that 'shrines were erected to historic inventors'.[50] Exceptionally skilled craftsmen could be honoured with a biography.[51] Had Jan van der Heyden been Chinese, he would doubtless have had a shrine erected in his name or have been the subject of a biography, too. But of material incentives for individual inventors such as those he enjoyed in Europe, there is no trace. If they existed in Ming or Qing China, evidence of this has, as yet, not come to light.[52]

Comparing Creativity in China and Europe

Settings and incentives formed part of the composite mixture from which technical innovations emerged. After 1500, Europe began to diverge from China in that both the settings and incentives in Europe showed an increasingly diverse combination of religious and secular elements, which in many ways had no equivalent in China at all. Moreover, resources, mental operations and design practices in the creation of innovations began to differ, too. Differences between Europe and China arose in the means by which new things were fashioned, in the mental steps by which ideas for inventions were turned into reality, and in the practical ways in which these were shaped into techniques or artefacts that actually worked.

These differences show up clearly in the nature and uses of technical drawings. The art of making technical drawings in China, as Chapter 3 described, had reached a high level of sophistication by the early fourteenth century, but techniques of graphic representation did not improve thereafter to the extent that illustrations could give a realistic image of machinery of any complexity. Illustrations of machines in China generally remained simpler and cruder than those in Europe. Peter Golas has remarked that Wang Zhen, author of the *Nongshu* of 1313, was a 'rarity among Chinese writers in technological matters in his intent that his drawings not only be accurate but that they serve, together with his

[49] Davids, *Rise and decline*, 414–416.
[50] Elvin, *Pattern*, 314.
[51] Hofmann, 'Biographer's view', esp. 296–297.
[52] Rowe, 'Political, social and economic factors', 35–36.

descriptions, as guides for the actual making [of] machines and utensils'. Technical illustrations usually had a didactic purpose or were primarily meant to be enjoyed as works of art. 'For most Chinese artists who did illustrations of machines... it probably never occurred to them that their drawings *could* or *should* be any more precise and accurate than they were'.[53] And because these images of machines did not reach a high degree of realism, he surmises, they did not 'invite [Chinese artists] using them for thinking about technological questions' either.[54]

In Europe, by contrast, the uses and styles of machine drawings kept evolving throughout the period between 1400 and 1700. Technical illustrations served more purposes than simply that of presenting devices to a broader public for didactic or aesthetic ends. Apart from presentational functions, which were their initial *raison d'être* in the Late Middle Ages, from the sixteenth century onwards, engineers' drawings of machines also helped the actual building of a mechanical device by serving as a means of communication with, on the one hand, the people who commissioned the project (for example, a princely ruler or a city government) and, on the other hand, the artisans (such as carpenters or blacksmiths) who constructed the machine on site[55] – a function intended by Wang Zhen but rarely realized in China. Furthermore, engineers in the fifteenth and sixteenth centuries began to collect drawings of their own projects and devices designed by others as a sort of personal archive, stored for future reference by themselves, colleagues, artisans or potential clients. Yet another function of drawings to emerge in this period, exemplified by the notebooks of Taccola and Da Vinci, was the use of drawings as a means to investigate on paper by devising a number of variations of a given device and thinking about the possible effects on motions and practicability. In a later stage, 'thinking' on paper sometimes took the form of some theoretical reflection on the properties of machines with the help of geometrical analysis.[56]

Changes in uses were intertwined with changes in styles in drawings. The styles of machine drawings evolved from 'logical schemas built of icons' to 'depictions of appearance' – from a picture that gave a general idea of the main parts and the general arrangement of a mechanical

53 Golas, 'Technological illustration', 50, 54–58.

54 Golas, '"Like obtaining a great treasure"', 593.

55 Popplow, 'Why draw pictures', 28–36.

56 Popplow, 'Why draw pictures', 36–48, McGee, 'Origins', 73–82, Galuzzi, 'Portraits of machines', 90, Long, 'Francesco di Giorgio', 138–139.

device, to a drawing that showed how machines actually looked.[57] Drawings depicted quite accurately the overall appearance of a machine from a single point of view and gave detailed views of separate parts, if need be by artificial views such as cutaways or exploded views.[58]

To some extent, these changes in pictorial style built on advances in geometrical techniques realized by master masons and architects in the High Middle Ages. They were also made possible by the recovery, reinvention and refinement of techniques first introduced in Antiquity, such as projection in perspective.[59] Theoretical reflections on the properties of machines were stimulated by the rediscovery of writings on mechanics from Hellenistic and Arab sources, such as those by Archimedes and Hero of Alexandria. Juan de Herrera, for example, kept copies of their works in his private library.[60] Initially, engineers, architects, master gun-makers and other technical experts primarily ventured on changes in drawing style in order to be better able to communicate with potential clients: in that context, 'depictions of appearances' were more effective then schematic 'icons', which could not be understood by a layman. But the growing social status, autonomy and ability of these experts also gave them more room to explore new directions for their own sake – including trying out new technical solutions on paper.[61] Moreover, authors of drawings of machines and mechanical devices 'showed a very strong spirit of emulation of the humanists' in 'recovering classic texts on technology and architecture', by stressing the importance of 'professional experience and drawing skills' in addition to 'textual restoration' and 'philologic methods', Paolo Galuzzi observed.[62] If Europe in the early modern period made greater strides in 'mechanical technologies' than China, as Bozhong Li suggests,[63] this may well have to do with these differences in the uses and styles of technical drawings, which reflect underlying differences in relations between technical experts and society at large.

Comparing technological creativity in China and Europe in the very long run, Joel Mokyr has advanced the thesis that more so than in Europe, technological progress in China depended on support from the

---

[57] McGee, 'Origins', 59, 63–66, 83.

[58] Lefêvre, 'Introduction', 5–6, McGee, 'Origins', 72, 82–83, Galuzzi, 'Portraits of machines', 76, Leng, 'Social character', 104–105.

[59] Lefêvre, 'Introduction', 6, Peiffer, 'Projections', 245, 254–267.

[60] Popplow, 'Why draw pictures', 43, López Piñero, *Ciencia*, 245.

[61] McGee, 'Origins', 83–84, Leng, 'Social character', 88–106.

[62] Galuzzi, 'Portraits of machines', 54.

[63] Li, *Agricultural development*, 168.

state. Technological change was generated in large part by public officials and the central government. And this dependence on the state became fatal once the attitudes of the state became less sympathetic. When the Chinese state lost interest in promoting technological change, progress could not be sustained, because 'there were no substitutes for the state'. In Europe, by contrast, technological change could be sustained in the long run, 'precisely because [it] was private in nature and took place in a decentralized, politically competitive setting'.[64] This setting was favourable to technical innovation both because it facilitated the emergence of new inventions and because it reduced the chances of easy elimination of innovative ideas. Although this picture somewhat exaggerates the actual contrast between China and Europe, to my mind, it nevertheless captures an essential truth. It exaggerates in the sense that it overemphasizes the role of the state in technological change in China and underestimates its weight in Europe. Substitutes for the state in China were not entirely absent: Buddhist monasteries, as we have seen, were instrumental in promoting technical change under the Tang and private entrepreneurs came to the fore under the Song and the late Ming. In Europe, the role of the state was actually less 'secondary and passive' than Mokyr suggests, and its contribution was not almost entirely restricted to 'nonmilitary technology'.[65] Governments often made important contributions to the improvement of hydraulic infrastructure, for example, and in the early modern era, they became increasingly active in promoting technological change by granting awards for the introduction of technical inventions, as we have seen. By and large, however, the distinction between China and Europe holds.

Competition in Europe from the High Middle Ages was more than rivalry between a large number of states. Many other sorts of actors also participated in the race for wealth and political power, ranging from popes, bishops and religious orders via princes, dukes and noblemen to cities, merchants and craft guilds. The decentralization of the competitive setting, where innovation could thrive, went well below the level of states. But Europe did not merely differ from the Chinese Empire in the sheer multitude of actors involved. 'Competition' as such was valued differently, too, which eventually had far-reaching consequences for the development of knowledge.

---

64 Mokyr, *Lever of riches*, 237–238.
65 Mokyr, *Lever of riches*, 233.

Learning in China, according to Nathan Sivin, Geoffrey Lloyd and Shigeru Nakayama, normally emphasized consensus.[66] Yet polemics between schools of thought, or between individual scholars, were not unknown. Scholars engaged in 'evidential learning', which flourished in private academies in Jiangnan during the early Qing, fought out priority disputes and had heated debates about interpretations of texts.[67] As a rule, however, Chinese scholars sought harmony and reconciliation between differing systems rather than disputation and verdicts by powers of persuasion. Chinese learning concentrated on documenting, cataloguing and finding correlations. Yet, although Chinese scholars *did* show interest in uncovering regularities and general patterns, they normally did not think in terms of 'laws of nature'. Ideas akin to the concept of such laws were not wholly absent in Chinese philosophy, but examples of writings expressing such notions are few and far between. In the Chinese context, this way of thinking always remained a minority viewpoint.[68] Scholarship and technology in China hardly connected at an analytical level. To be sure, scholar-officials since the Song *did* produce writings about the state of the art in different areas of technology, as we have seen. Except for the Mohists, however, who reached the peak of their influence in the late Warring States period, as Derek Bodde observed, no 'sizable and long-lived group among the classical schools of Chinese philosophy [showed] serious analytical interest in logic, natural science and technology'.[69] And engineers or craftsmen, for their part, did not record their reflections on their practice either.

Europe developed in a different way. The tradition of learning cultivated in European universities from the High Middle Ages onwards was characterized by what Nakayama calls 'the disputative approach': 'the skillful use of logic to persuade others to accept theories and interpretations'. 'Schoolmen', according to Nakayama, 'approached nature with the eyes of lawyers'.[70] This disputative approach was in itself not entirely new. Lloyd and Sivin observed that 'ancient Greek culture', much more than Chinese culture, encouraged 'disagreement and disputation in natural philosophy and science as in every other field'. Persuasion by rhetoric was a favourite

---

[66] Nakayama, *Academic and scientific traditions*, 54–55, 81, Sivin and Lloyd, *Way and the word*, 189–193, 247–248.

[67] Elman, *From philosophy to philology*, 200–202, 221–225.

[68] Needham, *Science and civilisation*, vol. II, 518–583, esp. 544, 582, Bodde, *Chinese thought*, 344, Nakayama, *Academic and scientific traditions*, 72.

[69] Bodde, *Chinese thought*, 354, 360–361.

[70] Nakayama, *Academic and scientific traditions*, 54, 71–72.

tool for making a reputation – and even for earning a livelihood. Above all, it was a means of competition.[71] And this very method of 'dialectical reasoning characteristic of ancient Greek philosophy' was adopted in the study of law, theology, philosophy and other subjects in European universities in the Middle Ages.[72]

Moreover, unlike China, Europe *did* develop a concept of 'laws' pertaining to physical nature. While the notion of a physical law of nature rarely appeared among ancient writers, the idea of 'law' or 'rule' occurred more than once in studies on optics, astronomy or mathematics from the thirteenth century, and turned into a fully-fledged concept of 'laws of nature' in the early modern period.[73]

The meaning of this development can be illuminated with the help of a distinction introduced by Joel Mokyr in his analysis of the growth of the knowledge economy. Mokyr distinguishes between two types of 'useful' knowledge: 'prescriptive' knowledge and 'propositional' knowledge. Prescriptive knowledge is 'how' knowledge: it consists of techniques, i.e. 'executable instructions or recipes' for ways to manipulate nature. Propositional knowledge – 'what'-knowledge – encompasses all knowledge about natural phenomena and regularities. It can assume two forms, Mokyr explains: one 'is the observation, classification, measurement, and cataloguing of natural phenomena', the other is 'the establishment of regularities, principles and "natural laws" that govern these phenomena and allow us to make sense of them'.[74] What was especially new in late medieval and early modern Europe was the growth of this latter form of propositional knowledge. After 1400, as we have seen, the circulation of knowledge via texts and images in script and print, composed by architects, engineers, artists, master gun-makers and other artisans for a variety of audiences, strikingly increased, while the 'spirit of emulation of the humanists' among these groups grew stronger as well. In addition, by the later sixteenth century, technical experts, with the help of 'propositional knowledge' on mechanics, more frequently included theoretical reflections on the properties of machines in their drawings. In contrast with

71 Sivin and Lloyd, *Way and the word*, 247–249; the role of the 'adversarial approach' in the making of the scientific culture of the West is a subject of a debate that does not concern us here, see e.g. Huff, *Rise of early modern science*, esp. 269, and Gaukroger, *Emergence*, 33–34, 40–41.

72 Berman, *Law and revolution*, 132–133.

73 Daston, 'Nature of nature', 160–161, Wilson, 'From limits to laws', 15, 21, Oakley, *Natural law, laws of nature*, 42.

74 Mokyr, *Gifts of Athena*, 4–5, 10.

China, prescriptive and propositional knowledge in Europe thus gradually became more intertwined.

What made these cognitive changes in Europe possible? Religious forces got the process going and, for a long time and to a large degree, continued to drive it. The catalyst was the 'Papal Revolution' in the eleventh and twelfth centuries, as discussed in Chapter 2. The Papal Revolution essentially consisted of the establishment of the supremacy of the papacy over the entire Church and the emancipation of the clergy from control by secular authorities; the revolution went together with the creation of a nuclear papal bureaucracy and the constitution of the 'rule of law' by the formation of a new legal system.[75] The papacy, according to Harold Berman, created 'the first modern Western' governmental institutions – a professional judiciary, a professional treasury, a chancery – and 'the first modern Western legal system' – the 'canon law' of the Roman Catholic Church – all of which served as models for the secular systems of government and law that arose in the following centuries. The church itself was now conceived as 'an independent, visible, corporate, legal structure'.[76] Essential to the structuring of canon law, as well as to the study of theology, which gave it some of its basic concepts, such as crime and punishment, was 'the dialectical method of scholastic philosophy'. Berman and others have called this 'method of analysis and synthesis of texts' 'the prototype of modern Western science', because 'it took the customs and the rules as data and adduced from the data the regularities – the "laws" – that explained them'.[77] The rise of canon law thus promoted the growth of a form of 'propositional' knowledge – 'the establishment of regularities, principles and "natural laws" that govern these phenomena and allow us to make sense of them' – that is regarded as the hallmark of European, as contrasted with Chinese, thought. The evolution of these notions into a concept of (physical) 'laws of nature' did not occur in any direct, linear way, however. That transformation largely took place in the sixteenth and seventeenth centuries and was closely linked with contemporary upheavals in politics, religion and philosophy.[78]

---

[75] Berman, *Law and revolution*, 2, 85–119, 520–537.

[76] Berman, *Law and revolution*, 521, 529–531.

[77] Berman, *Law and revolution*, 529, cf. Shapiro, *Culture of fact*, chapter 1. Benjamin Nelson located the decisive transformation in Europe in the twelfth and thirteenth centuries and saw these connected with the increased importance of ideas of 'universality' in the Western Christian world, see Nelson, 'Sciences and civilisations', 463–467, 475–476.

[78] Daston, 'Nature of nature', 160–161, Wilson, 'From limits to laws', Oakley, *Natural law, laws of nature*, chapter 2.

Teaching and studying law was the primary purpose of the universities founded in Europe from the end of the eleventh century onwards, starting in Bologna in 1087 and soon imitated in many cities in other parts of Italy and north of the Alps. Most of the newly-created universities formally came under the control of a bishop, albeit with a greater degree of autonomy than cathedral schools,[79] and they were initially largely staffed by clerics. Throughout the Late Middle Ages and the early modern period, religious organizations, as we have seen in the previous chapters, continued to make substantial contributions to the formation of human capital, via universities and other schools and institutions, and to the circulation of technical knowledge. As patrons of large projects in building and hydraulics, religious institutions provided a friendly ambience for innovations by architects, engineers, artists and artisans. As important providers of technical and vocational schooling in Spain, Italy, France or Central Europe after 1500, they moreover helped to forge concrete links between 'propositional' and 'prescriptive' knowledge.

The interplay between competition, cognitive change and religious institutions can be illustrated, in a concentrated form, by once again taking a look at the technology of river control in China and Europe, more specifically in the basins of the Yellow River and the Yangzi and the coastal plains of Northern Italy. In the Late Middle Ages and early modern period, these areas in China and Italy were highly vulnerable to recurrent calamities caused by the unruly behaviour of rivers forcing their way in from a mountainous hinterland.[80] Although the Po, the Adige, the Piave, the Reno and other rivers did not bring down as much silt as the Yellow River or the Yangzi in China, nor showed such violent fluctuations over the year, sedimentation and torrential flows nevertheless caused serious problems, especially on the plains of the Veneto and the Emilia-Romagna, where numerous larger and smaller streams converge towards the Adriatic Sea.[81]

The key difference between the Chinese and Italian regions was the emergence of a theory on river hydraulics in Northern Italy. This 'theoretical turn', which began in the Papal States in the 1620s and reached the Venetian Republic a few decades later, essentially consisted, as Cesare

---

79 Berman, *Law and revolution*, 123–127.

80 Davids, 'River control', 61–64.

81 Ciriacono, *Acque e agricoltura*, 138–139, 162–164, 196–201, Cazzola, 'Le bonifiche cinquecentesche nella valle del Po', 15–17, Maffioli, *Out of Galileo*, 24, 42–43, 156–158, 347, 371.

Maffioli put it, of reshaping the existing tradition of fluvial hydraulics 'in a geometric fashion, around the basic concept of velocity', in order to obtain more reliable knowledge about the motion of waters in rivers.[82] Its founding father was a Benedictine monk who taught mathematics at Pisa and Rome, Benedetto Castelli. Castelli's treatise *Della misura dell'acque correnti*, with its companion piece, *Demostrazioni geometriche della misura dell'acque correnti*, published in Rome in 1628, was the first to approach the phenomenon of the behaviour of rivers with the full panoply of definitions, suppositions, propositions and demonstrations, which was known as the 'geometric' way.[83] In the following decades, and by the combined efforts of a host of other Italian scholars, including Evangelista Torricelli, Geminiano Montanari, Domenico Guglielmini, Guido Grandi, Bernardino Zendrini and Giovanni Poleni, this 'geometrical' approach to river hydraulics was extended into an elaborate corpus of general concepts, principles and laws relating to the motion of waters. Maffioli has observed that in around 1700, many Italian contributions to the European scientific debate 'were directly or indirectly related to the science of waters'.[84]

River hydraulics in the Chinese regions did not go through the same theoretical turn as in Northern Italy. Why did they not make a similar 'cognitive leap'? Part of the explanation may lie in the fact that one basic cognitive element required for making this turn, namely the availability of 'deductive geometry in the Western sense', was simply lacking. 'Chinese mathematics', after all, is said to have been 'rather focused on arithmetical and algebraic procedures'.[85] Yet, there must have been more to the matter than the mere presence, or absence, of particular intellectual tools. The supply of such tools was not an invariable given. Euclidean geometry and various European innovations in methods of calculation *did* reach China through the intermediary of the Jesuits in the course of the seventeenth century.[86] The study of river hydraulics in Qing China thus might have drawn upon similar resources to those available in Europe. To explain the differences in development, other circumstances have be taken into account as well.

---

[82] Maffioli, *Out of Galileo*, 419–420; earlier work on obstacles in river courses had been done by Leonard da Vinci.

[83] Maffioli, *Out of Galileo*, 41, 45–51.

[84] Maffioli, *Out of Galileo*, 14, and Tables 1.1. and 1.2.

[85] Cohen, *The Scientific Revolution*, 440, following Needham, *The Grand Titration*, 44.

[86] Jami, 'Learning mathematical sciences', 229–231, Elman, *From philosophy to philology*, 180–184.

Now, one of the striking features of the Chinese case is that both 'prescriptive' and 'propositional' knowledge about fluvial hydraulics was apparently entirely produced *within* the central bureaucracy that was concerned with controlling the Yellow River. This elaborate bureaucratic structure *did* allow the emergence of divergent views about ways to solve the problems of managing the Yellow River – witness the intense debates and dramatic shifts in policy in the time of Pan Jixun at the end of the sixteenth century, or during the tenure of Jin Fu as Director General of the Conservation of the Yellow River in the 1680s.[87] However, it did not result in the rise of new forms of propositional knowledge, as in Northern Italy. Outside this central institution, there were no complementary, or rival, sites of knowledge production and distribution about river hydraulics. The study of river hydraulics in China still could have witnessed a similar cognitive leap to that seen in Northern Italy, if the knowledge accumulated by experience at a local level had been reshaped into a system of more abstract, generalized 'science of waters', for example in the context of the private academies in the Lower Yangzi area, which formed the heart of the movement of 'evidential learning'. Yet, there is no indication that scholars in these academies made an effort to develop a mathematical approach to the field of fluvial hydraulics either. They were not interested in applying their knowledge to a practical concern such as the study of rivers.[88]

The breakthrough in Italy in the second quarter of the seventeenth century was to some degree related to the existing tradition of hydraulic engineering, which had flourished since the Renaissance. Benedetto Castelli and his followers could build on the accumulated knowledge of generations of hydraulic practitioners. But this was only part of the story; if it had been the whole story, it would be very puzzling indeed that a similar leap was not accomplished in China. Another part of the story was that from the start, the new approach was also clearly contrasted with the tradition of the local practical experts (*proti* and *periti*). Champions of the 'science of waters' claimed that their 'geometrical way of thinking' would generate more reliable, and therefore more useful, knowledge than the empirical approach taken by the *proti* and *periti*. The origin of this theoretical turn should, according to Maffioli, be sought in the extension of

[87] Vermeer, 'P'an Chi-hsün's solutions', Dodgen, *Controlling the Dragon*, 18–22, Elvin, *Retreat of the elephants*, 135–140, Strassberg, *The world of K'ung Shang-Jen*, 118–120, 350–351.

[88] Cf. Jami, 'Learning mathematical sciences', 227.

the Galileo's mathematical approach to nature, which was initially developed to study the motion of solid bodies, to the domain of the motion of waters. Galileo's 'new science' thus acquired a wider field of application. And Castelli was in a unique position to forge the link, because he was a hydraulic consultant at the papal court, a mathematics professor as well as one of the closest collaborators of Galileo himself.[89] This does not imply that the new departure in river hydraulics consisted entirely of the transplantation of Galilean concepts and methods. Later generations of scholars made important additional contributions to the field by applying new intellectual tools, such as the calculus, and by refining the use of experiments.[90] The key change, in retrospect, was the fusion of insights from two different traditions of knowledge, which eventually provided the field of river hydraulics with a more extended, varied set of propositions about nature than the *proti* and *periti* alone would have been able to muster.

A characteristic feature of the North Italian setting was the combination of the relative precocity of the bureaucratic organization of river management with high diversity in the social and institutional basis of knowledge on fluvial hydraulics. Hydraulic administrators appeared relatively early, but the production and distribution of knowledge on the subject was fairly dispersed. A monopoly in this field did not exist. At an early date, hydraulic offices were to be found in several states and cities in Northern Italy. In Venice, for instance, a magistracy for the supervision of canals was instituted in 1224. A *Magistrato all' Acqua*, responsible for handling all hydraulic problems, was established in 1501.[91] Another office, charged with taking care of the River Adige, was erected in 1677, with branch offices in Verona and Padua. The managers of these boards, who were members of the Venetian patriciate, could call on a small staff of technical experts, called *proti*.[92] In the late sixteenth century, experts from Venice were repeatedly called upon to advise on the solution of hydraulic problems in the environs of Ferrara.[93] Bologna also founded a hydraulic board, called the '*Assunti* to the waters'. These *Assunti*, recruited from the Senate, likewise received assistance from a staff of practical experts, the

89 Maffioli, *Out of Galileo*, 37–51, 418–23.
90 Maffioli, *Out of Galileo*, parts III and IV.
91 Ciriacono, *Acque e agricoltura*, 140, Lane, *Venice*, 16.
92 Maffioli, *Out of Galileo*, 276–277.
93 Fiocca, 'Regolamentazione', 139–141.

periti. Another post, that of superintendent of the waters around the city, was established in 1686.[94]

Knowledge of river hydraulics did not remain confined to these special offices concerned with water control, however. Other sites of knowledge production and transmission emerged at Jesuit colleges and universities in the seventeenth century. It was at these institutes for higher learning, which were not dependent on the old-established hydraulic offices, that the theoretical turn in fluvial hydraulics first occurred. *Proti* and *periti* were naturally neither overjoyed by its appearance nor quickly convinced of its use. The rise of these rival centres of knowledge was apparently the outcome of three parallel developments. First, one of the driving forces was the growing competition for students between the university of Bologna and Jesuit colleges, as well as other institutes of higher learning in the Papal States. Faced with a serious crisis caused by the diminishing attractiveness of the local university to foreign students and the local nobility, the municipal government of Bologna (as supervisor of this institution) actively welcomed innovation in the curriculum in the later seventeenth century. Second, at the same time, the new approach in hydraulics could also make headway at the state university of the Venetian Republic in Padua, thanks to the support of the Venetian patricians, who set great store on the knowledge of the newly risen 'scientists of the waters' as an alternative source of expertise to the traditional lore of technical practitioners.[95] Third, for scholars themselves, employment at a university became more attractive as an avenue to make a career in science, as opportunities for patronage from princely courts declined. In all these respects, Italian scholars after Galileo truly struck out into a new direction.[96]

Religion, Technical Innovation and the Little Divergence

In the course of time, technological creativity varied not only between China and Europe, but also within Europe. The differences that arose within Europe were not as big as the gap that opened up between China and Europe from the eighteenth century. The disparities between developed for a while between c.1100 and 1800 between regions that achieved

94 Maffioli, *Out of Galileo*, 172, 181–182.

95 Maffioli, *Out of Galileo*, 132–135, 243–249, 274–277, Buonora, 'Cartografia e idraulica del Tevere', 178.

96 Maffioli, *Out of Galileo*, chapter 6 and 422–428.

technological leadership (first Northern Italy, Southern Germany and the Southern Netherlands, then the Dutch Republic and, finally, Britain)[97] and other areas of the Continent were never so great that they could not been bridged within the time span of a century. Even when in the late eighteenth century, Britain realized a breakthrough of revolutionary proportions, resulting in a fundamental shift in the structure of the economy at large, the divergence did not increase to such a extent that other European countries were unable to catch up with her before the end of the nineteenth century; many of them, in fact, did so within a remarkably short time. To what degree can these variations in technological creativity *within* Europe be explained by variations in religious context? This is the key question that will be addressed in the second part of this chapter.

I will address this question in three steps. First of all, I will examine when and to what extent different regions in Europe went through a more or less common trajectory in technological creativity. How long did their 'shared' experience last? The second step concerns divergences between regions in the creation of technical innovations: when and how did these divergences actually occur and in what ways were they expressed? The answers to these queries lead up to the final step: inquiring about the extent to which these emerging differences in technological creativity can be explained by variations in religious contexts.

Common Trajectories

Chapter 2 showed that with regard to human capital, the extant evidence does not confirm the common assumption that Protestant regions of Europe had a lead over Catholic ones, nor does it corroborate conventional assumptions about consistently growing differences between north-western and southern Europe. In human capital formation, different regions of Europe largely shared the same experience throughout the Late Middle Ages and early modern period. The Protestant and Catholic Reformations from the sixteenth century were, on balance, not an obstructing factor for the circulation of technical knowledge either. Although at times, these religious transformations led to massive waves of 'confessional migration' of skilled people and in many places had an influence on admission policies for master-craftsmen, the actual impact of these changes on regional or local technological development, either in a positive or a negative sense,

[97] Davids, 'Shifts of technological leadership', 338–341.

was not very marked. Chapter 3 concluded that the 'Little Divergence' in economic performance between areas in North-western Europe and other regions of the Continent, which often happened to be Catholic, can no more be ascribed to some retarding impact on the part of the Tridentine Church than in the case of human capital formation.

Shared experiences in human capital formation and circulation of knowledge between regions in Europe can be illustrated by looking at connections between the development of knowledge on river hydraulics in the Dutch Republic and Northern Italy before 1800. In the Netherlands, as in China, nearly all writings on this subject composed before the 1770s were either of the 'prescriptive' type (instructions on how to deal with specific hydraulic problems) or of the 'propositional' sort: observations, classifications and measurements of natural phenomena. By the end of the seventeenth century, for instance, it had become normal practice among surveyors or engineers to cast proposals to solve the problem of the distribution of the water of the Rhine in the form of written memoranda, often accompanied with maps, which were based on rules derived from experience, as well as a series of soundings and careful observations of the situation on the spot.[98] From the 1720s onwards, surveyor Nicolaas Cruquius made a major improvement to this approach by grounding every proposal, advice or statement about hydraulic matters in an extensive database of measurements of hydraulic variables, and making as much use as possible of cartographic aids to record and analyse the resulting data. Other experts soon followed his example.[99]

In contrast with China, however, the corpus of knowledge on river hydraulics in the Dutch Republic later grew to include 'propositions' of Mokyr's second form, namely statements about 'regularities, principles and "natural laws"'. Scientific societies began to stimulate thinking about the theoretical aspects of hydraulics from the 1770s onwards. The first volume of transactions published by the *Bataafsch Genootschap der Proefondervindelyke Wijsbegeerte* in 1774, for example, opened with a treatise running to over 200 pages by a medical doctor, Lambertus Bicker, about the basic principles of river management and their application in the case of the Dutch Republic.[100] A lively public debate on issues related to river control arose, in which the participants were keen to bolster their

98 Van de Ven, *Aan de wieg van Rijkswaterstaat*, 64–102.
99 Van den Brink, *'In een opslag van het oog'*, chapters 1 and 4.
100 Bicker, 'Rivierkundige grondwaarheden'.

positions with theoretical arguments. Education about hydraulic matters was no longer confined to on-the-job training or to the individual perusal of the writings of famous predecessors. Johan Frederik Hennert, professor of mathematics, astronomy and physics at the University of Utrecht, began to offer 'public lectures on the course of rivers' from the 1780s onwards. At the University of Leiden, hydraulics was taught by Jan Frederik van Beeck Calkoen, who held the chair of natural philosophy from 1799.[101] Pupils at the privately endowed *Fundatie van Renswoude* in Delft who chose to become hydraulic engineers received both training 'on the job' and a thorough grounding in mathematics and physics from teachers at the institute itself.[102]

This cognitive change in the Netherlands owed much to the growing influence of the Italian model. Italian influences were transmitted both through the mobility of people and by the spread of knowledge stored in printed works. The Italian engineer and founder of the Institute of Arts and Sciences in Bologna, Luigi Fernando Marsigli, for example, who was a long-time correspondent of professor Herman Boerhaave in Leiden, resided in Holland for over a year, in 1722 and 1723. During his stay, he made several trips to view the hydraulic 'sights' in the coastal provinces, in the company of Boerhaave and Cruquius. His visit was to inspire Cruquius to introduce curves of equal depths in river maps about 1730, which has been a normal feature in the cartography of rivers in the Dutch Republic ever since.[103] Another Italian hydraulic scientist of note, Paolo Frisi, who held the chair of mathematics and philosophy in Milan, kept up enduring contacts with Dutch colleagues after a journey to the Netherlands in 1766.[104] A Dutch translation of a short treatise by Frisi on the division and confluence of rivers, which was partly based on the theoretical insights and practical experiences gained by experts in Italy as a result of the long debate on the regulation of the River Reno, appeared in 1773 in the transactions of the oldest learned society in Holland, the *Hollandsche Maatschappij der Wetenschappen*.[105]

Dutch academics and engineers also learned about advances in hydraulic phenomena in Italy simply by studying their books. Willem Jacob

[101] Maffioli, 'Italian hydraulics', 250, *Nieuw Nederlandsch Biografisch Woordenboek*, vol. IX, 123–124.

[102] De Booy and Engel, *Van erfenis tot studiebeurs*, 72–81, 105–112.

[103] Van den Brink, *'In een opslag van het oog'*, 59, McConnell, 'A profitable visit', *passim*.

[104] Maffioli, 'Italian scientists', 257–258.

[105] Frisi, 'Berigt aan de Hollandsche Maatschappij der Weetenschappen'.

's Gravesande, for instance, a professor of mathematics and astronomy in Leiden, who influenced Cruquius's comprehensive, quantitative and highly systematic approach to problems of river improvement,[106] was thoroughly acquainted with the work of, among others, Domenico Guglielmini, Guido Grandi and Giovanni Poleni. Theoretical insights and methods developed by Guglielmini and Poleni served as a source of inspiration to Hennert and Christiaan Brunings, Inspector-General of the Rivers, in the 1780s.[107] When his regular bookseller failed him, Brunings did his utmost best to get hold of Italian publications about hydraulics through his connections in the literary world.[108] Circulation of knowledge and human capital formation thus connected regions in Europe across religious boundaries.

Divergences in Creativity in Europe

When and how, then, did technological creativity start to differ between regions within Europe? When and how did divergences in creativity occur, and in what ways were they expressed? Between the High Middle Ages and the Industrial Revolution, Europe witnessed the ascent and eclipse of technological leaders several times. The rise and decline of 'technological leadership' between geographical areas became a regular occurrence. 'Technological leadership' meant that a given country, region, town or cluster of towns played an initiating role in the development of new technologies in a wide variety of fields for a prolonged period of time. Technological leadership shifted from one territory to another.

The life-cycles of technological frontrunners showed a more or less fixed pattern. Before starting on a lengthy technological advance, a future leading region, towns or cluster of towns first developed into the hub of a widespread trading network. Usually, the scope of the technological advance was fairly narrow at first, but once the domination over a large trade network had been firmly established, it broadened in many ways. No centre of leadership started from scratch. Each new centre partly built on the achievements of its precursors. Southern Germany borrowed technical knowledge from Northern Italy, Flanders and Brabant borrowed from Italy and Southern Germany, the Dutch Republic borrowed from all three, and Britain, in its turn, borrowed from the Dutch and their predecessors. The circulation of knowledge

---

106 Van den Brink, *'In een opslag van het oog'*, 13–18, 24–25.

107 Maffioli, 'Italian scientists', 238, 252–3, Van Schaik, *Christiaan Brunings*, 12–16, 56–58, Brunings, 'Antwoord op de vraag'.

108 Van Schaik, *Christiaan Brunings*, 77.

thus contributed to the accumulation of know-how from one centre of leadership to another.[109] Although a technological frontrunner was often unable to retain a competitive edge in all of the fields where it had achieved initial primacy, it could still, for a while, offset the loss of the original strongholds by building new ones. Frontrunners ventured on novel routes of technological advance. As the evolution of technology proceeded, its nature normally changed as well. While the initial achievements of technological leaders usually consisted more of improvements in physical productivity than of gains in quality, the order was often reversed later on. Advancements in the quality of production came to be prized more highly, or were more readily achieved, than increases in physical productivity. Eventually, however, substitutions of old sectors of growth by new ones became ever more rare. Leadership gradually shifted to the town, region or country that boasted a higher rate of innovation.[110]

Before 1800, all successive centres of technological leadership in Europe were situated in the central urban belt in the form of an arc running upwards from Northern Italy through Switzerland and Southern Germany, into the Rhineland and the Low Countries, curving across the North Sea, to end in the West Midlands in England.[111] At times, nodes of technological creativity also existed outside this central belt, for instance in Northwest France, the Lyon region, Catalonia, the Basque Country or Portugal. Yet, these places did not play an initiating role in the development of new technologies in such a wide variety of fields or for such a prolonged period of time as in regions like Venice, Flanders or the Dutch Republic. Their creativity was short-lived or concentrated in a specific area of technology, such as ship construction, silk manufacture, mining or the art of navigation.

Leadership in technology did not solely depend, of course, on dominance in an extended trading network and the inflow of knowledge from other regions. There were other effective forces and factors, too. The high rate of technological innovation in the Dutch Republic after c.1580, for instance, can in some measure be explained by the evolution of relative factor prices. The rise in labour costs and the fall in the price of capital that could be observed from the last quarter of the sixteenth century onwards presumably made it a rational course of action for Dutch entrepreneurs to adopt labour-saving inventions. The attitudes of institutional bodies outside

---

109 Davids, 'Shifts of technological leadership', 339–341.
110 Davids, 'Shifts of technological leadership', 342.
111 On this concept, see Davids and Lucassen, 'Introduction', 11–17.

the marketplace were relevant, too. Guilds, *neringen*,[112] city governments, provincial authorities, admiralties, army agencies, drainage boards, chartered companies and various other public or semi-public bodies which were involved in decision-making about innovation were often quite willing to support the adoption of new technologies. Yet, technological advancement cannot be fully explained by changes of factor prices or the attitudes of institutions. The evolution of relative factor prices as such did not condition the *nature* of the reaction by entrepreneurs nor the *direction* which technological innovation would take. The mere fact of a change in relative factor prices in itself did not determine whether entrepreneurs would react by adopting technical innovations and, if so, what sort of innovation actually would come about.[113] And even if it is true that institutional bodies outside the marketplace could encourage technical innovation, that still leaves the question open as to how new technologies kept being created in the first place for a prolonged period of time. What was the reason why the *supply* of technical inventions varied; why was there more technological creativity in some regions than in others?

For some historians, the answer to this question seems to be fairly straightforward. Conservative attitudes and/or an absence of appropriate incentives led to a lack of creativity, while willingness to accept the 'new' could encourage it – but in the early modern period, such attitudes were still in rather short supply. According to Robert Friedel, in the sixteenth century, only a tiny portion of the population in Europe was 'stimulated by new knowledge and new ideas', 'moved by new opportunities and discoveries' or 'inspired by new capabilities or new problems'. It was a rare phenomenon indeed to see 'the quest for novelty' becoming 'a widely recognized value of the culture' as a whole; the first society in which this happened was the Dutch Republic in the Golden Age. Normally, 'there were many forces resisting the unfamiliar and the innovative'. Often these forces were lurking in institutions and the habits of people, and sometimes they 'were readily apparent and imposing – the Reformation answered by the Counter-Reformation'.[114]

---

112 *Neringen* were organizations established by a town government, charged with regulating and controlling a particular branch of export industry. In contrast with guilds, *neringen* had no formal membership but included *all* persons active in a specific branch of production.

113 Davids, *Rise and decline*, 534–535, cf. Roy, 'Knowledge', 373.

114 Friedel, *Culture of improvement*, 146–147.

In his magisterial study on European perceptions of Western and non-Western material achievements, Michael Adas is quite specific about the failings, or strengths, of settings and incentives in different regions in Europe. Adas confidently states that most French, Portuguese or Spanish authors of missionary and travel accounts in the sixteenth and seventeenth centuries 'displayed ... limited interest in technology', because they were 'from aristocratic or bourgeois families', which 'in Iberia and France in particular ... harbored a pronounced disdain for artisans or those engaged in the "useful arts"'. Observers from societies where disdain for manual labour was less pronounced, namely England and the Netherlands, 'often had a good more to say about technology' in their travel accounts.[115] Equally telling is the silence with which other studies on early modern technology shift their focus from southern Europe to England after 1600. In a book on perceptions of 'the rise of the machine' in European culture' between 1450 and 1700, Jonathan Sawday takes as his main examples for the fifteenth and sixteenth centuries the Italian engineers Leonardo da Vinci, Domenico Fontana and Agostino Ramelli, the German physician, engineer and metallurgist Agricola and the French philosopher Michel Montaigne. In chapters on the seventeenth century, however, readers find themselves suddenly transplanted to England. These chapters principally discuss the vision of the poet John Milton and ideas of the 'mechanical philosophers' who followed the trail blazed by Francis Bacon.[116] It is apparently regarded as self-evident that from the seventeenth century, the heart of the story was to be found in England, not in Italy, nor France, nor any other region in Europe.

But do such smooth narratives withstand closer scrutiny? Can early modern history really be fitted into such neat, simple schemes, in which southern Europe stands for everything conservative, prejudiced and stagnant and northwest Europe, especially England and the Netherlands, for all the opposite qualities? I think that such a stylized picture neither gives a fair account of what happened in the southern regions of Europe nor credibly portrays what happened in England and the Netherlands.

Take Spain, for a start. Recent research on the history of technology in Spain has uncovered a wealth of technical achievements in the sixteenth century, particularly in areas such as mining, hydraulic engineering and navigation technology. It was not until some time after 1600 that

---

115 Adas, *Machines as the measure of men*, 28–29.
116 Sawday, *Engines of the imagination*, chapters 1, 2, 3, 6 and 7.

technological creativity declined.[117] The results of this fresh scholarship should finally put the Black Legend about the obscurity and backwardness of Spain to rest. Between 1478 and 1650, some 240 patent applications were submitted to the King of Castile or to the Councils of Castile and the Indies, of which 152 were actually approved.[118] One of the most prolific inventors was Jerónimo de Ayanz y Beaumont. Ayanz belonged to the very social class that, according to Adas, 'harbored a pronounced disdain for artisans or those engaged in the "useful arts" ': the old nobility. He was related to the kings of Navarre. He held various high military and administrative positions under Kings Philip II and Philip III, including the post of administrator-general of mines from 1597 till 1604, and in the last years before his death in 1613, he was active as a mining entrepreneur himself.[119] Among the dozens of Ayanz's inventions (many of which he patented in 1606) were an improved method for extracting silver in the mines of Potosí, new sorts of ovens for use in the metal industry, in the army and at home, improved kinds of human-, animal-, and wind- powered mills, new diving equipment, new types of pumps for use on ships and new devices for lifting water in mines, including a rudimentary steam engine. This steam engine was actually brought into operation in a silver mine in Guadalcanal, near Seville, in 1611.[120] Other productive inventors, such Pedro Juan de Laspanosa and Juan de Herrera, who were from mercantile or gentry stock, succeeded thanks to their technical abilities in gaining a considerable amount of power and wealth; Herrera, who served as architect and quartermaster under Philip II, had amassed a sizeable fortune by the time of his death in 1597.[121] The stories of these inventors can hardly serve as evidence of disdain for the manual arts among 'bourgeois and aristocratic families' in the Iberian Peninsula. Far from being an obscure backwater, sixteenth-century Spain was in fact the country in Europe where the idea that the 'Moderns' in many respects were superior to the 'Ancients' first began to take root in humanist circles – long before this happened in England or France. And one of the reasons for this was the remarkable advancement in technology realized from the late Middle Ages onwards.[122]

---

[117] López Piñero, *Ciencia*, 196–212, 240–279, García Tapia, *Tecnica y poder*, Goodman, *Power and penury*.
[118] García Tapia, *Tecnica y poder*, 195–201, Mateos Royo, 'State policy', 149–150.
[119] García Tapia, *Inventor navarro*, 23–33, 123–142, 247–258.
[120] García Tapia, *Inventor navarro*, 147–156, 181–205, 209–225, 258–259.
[121] García Tapia and Carillo Castillo, *Tecnología y imperio*, 57–58 62–73, 79–91, 102–111.
[122] Maravall, *Antiguos y modernos*, 16, 481–575, Cañizares-Esguerra, *Nature*, 14–23.

Neither do other regions in southern Europe easily fit the stylized picture. Northern Italy in the Late Middle Ages and the sixteenth century was at least as fertile a country for technical innovations as Spain. Between 1474 and 1600, for example, the Senate of Venice received more than a thousand proposals for new inventions and granted no fewer than 610 patents. The silk industry, which was one of the most important branches of manufacturing in the city, showed a constant capacity for adopting new techniques and products.[123] And Italian engineering, as we have seen, long remained a paragon for other countries in Europe, especially in fields such as building, fortification, warfare and river hydraulics. Moreover, the claim that authors of missionary accounts from southern European countries showed limited interest in technology because they were from 'aristocratic or bourgeois families', which 'harbored a pronounced disdain for artisans or those engaged in the "useful arts"' sits awkwardly with what is known about the most global of all missionary organizations in the early modern period, the Society of Jesus. Jesuit missionaries actually had a variety of social backgrounds. They were not all sons from bourgeois and aristocratic families. As for their geographical origins, most of them came from Portugal, Spain, Italy, France, Central Europe and the Spanish Netherlands.[124] In contrast with what Adas claims, Jesuits' accounts do not betray a lack of interest in technology or disdain for manual labour. Nor were these shock troops of the 'Counter-Reformation' as averse to innovation as Robert Friedel assumes. Jesuits were often accomplished craftsmen, as Chinese emperors unfailingly noticed, and they were able to sent detailed and well-informed reports about 'useful arts' in the world beyond Europe. Jesuits in eighteenth-century China were in fact so skilful in the manual arts themselves that they were able to construct any device the Emperor wanted, and to such a degree of complexity, as one of them noted in 1754, that it 'en Europe même, passerait pour une merveille, ou tout au moins pour un chef d'oeuvre de l' art'.[125]

Making England the centrepiece of the story of 'the rise of the machine' as early as the seventeenth century, on the other hand, is pretty premature. The choice of illustrations in Sawday's book nicely, though unwittingly, underscores the point. Most of the illustrations date from the period before 1600 and they are largely taken from machine books or

[123] Molà, *Silk industry*, xix, 189.

[124] On origins and motivations of Jesuits, see Clossey, *Salvation and globalization*, 119–135.

[125] Vissière (ed.), *Lettres*, 455 letter Amiot to De la Tour, October 17, 1754.

paintings produced in Germany, Italy and the Low Countries.[126] Actually, the number of patents annually granted in England did not exceed the numbers granted in the heyday of patenting in Venice or the Dutch Republic until after the middle of the eighteenth century.[127] Before that time, England did not yet enjoy a reputation 'for being particularly original or inventive'.[128] Although England may have boasted many skilled 'mechanics and technicians',[129] there was not much demand for their services abroad before the last decades of the eighteenth century.[130] What is more, for a long time, Englishmen continued to look up to the Netherlands as a paragon of technological creativity.[131] In his famous travel guide, *The Grand Tour*, Thomas Nugent stated that as late as 1749, 'there [was] no nation where the people apply themselves with more diligence to all manners of mechanical arts, than the inhabitants of the United Provinces'.[132]

Thus, even though it is true that technological leadership in Europe between the Late Middle Ages and the mid-eighteenth century roughly shifted in a northwest direction, moving from Northern Italy via Southern Germany and the Low Countries to England, and even though it is agreed that in the seventeenth century, Spain eventually *did* become less creative than before, it would be wide of the mark to claim that variations in technological creativity can be directly reduced to fundamental differences between 'South' and 'North', or between 'Catholicism' and 'Protestantism' as such, however tempting the use of such conventional, easily comprehensible, binary schemes may be. Variations in technological creativity within the central urban belt of Europe, and between regions in this central belt and outlying areas, such as the Iberian Peninsula, cannot simply be explained by the prevalence, or weakness, of conservative attitudes and the presence or absence of appropriate incentives, which are supposed to have been in some way related to the outlook of particular dominant classes or to some ingrained contrast between the Reformation and the Counter-Reformation.

---

126 See the list in Sawday, *Engines of the imagination*, xi–xiii.

127 Davids, 'Shifts in technological leadership', 347–348, Molà, *Silk industry*, 186–189, Mandich, 'Le privative industriali Veneziane (1450–1550)', 537–547, MacLeod, *Inventing*, 150.

128 Mokyr, *Lever of riches*, 240.

129 Mokyr, *Enlightened economy*, 107–111.

130 Cf. Jeremy, 'Damming the flood'.

131 Davids, *Rise and decline*, chapters 4 and 5.

132 Nugent, *The Grand Tour*, I, 32.

Technological Creativity and Religious Contexts

This brings us to the third, and final, step in the analysis of variations in technological creativity *within* Europe: inquiring about the extent to which these emerging differences in technological creativity can be explained by variations in religious contexts. Religious variables, I will argue, *did* indeed affect technological creativity, but in more indirect, subtle ways and at a later moment in the technological development of Europe than conventional, binary schemes of 'North' versus 'South' or 'Catholic' versus 'Protestant' would suggest. Despite the predilections of economists and sociologists to the contrary, we would be better to cast aside our Weber-inspired stereotypes.

The fact that in the seventeenth and early eighteenth centuries, technological creativity in predominantly 'Catholic' regions in 'southern' Europe, such as Venice, Spain or France, lagged behind that in regions in 'northwest' Europe, such as the Dutch Republic and, later, England, where the Protestant Reformation gained the upper hand, cannot be ascribed to changes in dominant religious values, practices or institutions. Why should a religious environment that was evidently favourable to technical innovation before 1600 suddenly become inauspicious for technological creativity thereafter? Admittedly, the Catholic Reformation began to leave an ever stronger mark on these countries from the end of the sixteenth century onwards, but there was nothing in this religious reform movement per se that discouraged technological innovation. In fact, Catholic religious organizations made a major contribution to human capital formation in Italy, Spain, France, the Southern Netherlands and parts of Germany in the seventeenth and eighteenth centuries, and as we have seen in previous chapters, the Catholic Reformation boosted the circulation of technical knowledge in important ways.

To be sure, religious diversity within Catholic territories decreased between c.1580 and 1700. Dissenting communities of evangelicals, Anabaptists and millenarians in Venice, which had been quite strong in the mid-sixteenth century, had, as a result of repression by the Inquisition and the revival of popular forms of Catholic piety, almost vanished by the 1580s.[133] In Spain, hundreds of thousands of Moriscos were forced to leave the country after 1609. France expelled a large part of its Huguenot population in the 1680s and 1690s. These large-scale movements of forced

[133] Martin, *Venice's hidden enemies*, 15–16, 56–64, 211–215, 230–231.

confessional migration *did* lead to a temporary loss of human capital and in a few places – such as in the environs of Valencia – to some technological regression as well.[134] But there is no indication that the departure of these religious minorities dealt a serious blow to technological creativity in their country of origin. Fresh research on the specializations of Moriscos has cast doubt on the thesis that their expulsion implied a serious technological loss for Spain.[135] The most vocal champion of the role of Protestantism in science and technology in history, Isidor Thorner, does not present any evidence that 'ascetic Protestants' in Europe produced more 'inventions' than Catholics.[136] And even though quite a few champions of technological improvement in late sixteenth-century France were 'inspired by evangelical ideas',[137] there is no proof that Huguenots at that time, or later, were technologically more creative than Catholic subjects of the King. Admittedly, Moriscos or Huguenots sometimes contributed to technological advancement in their countries of destination, like the Maghreb, England or Switzerland, but this does not demonstrate that they were particularly creative in their country of origin. It only shows that there were differences in levels of technological development between countries, which could in some measure be reduced if skilled migrants helped to raise the level in the less advanced regions. The contributions of Moriscos and Huguenots were much less remarkable in host countries that were technologically more advanced, such as Tunisia or the Dutch Republic.[138]

The slowdown, or lag, in technical innovation in countries like the Venetian Republic, Spain and France can more readily be explained by political, social or economic changes than by religious factors. Changes in political, social or economic conditions adversely affected incentives and settings for technological creativity. From the second quarter of the seventeenth century, The Venetian economy underwent a marked structural shift from sectors manufacturing goods for exports to sectors servicing local or regional markets. This shift was connected to the loss of competitiveness of Venetian industries in international markets. Demand for Venetian products in distant markets declined. On the other hand, Venice

---

134 Scoville, *Persecution of the Huguenots*, Scoville,'Minority migration', Scoville, 'Huguenots and the diffusion of technology', Glick, *Irrigation and society*.

135 Glick, 'Moriscos y marranos', 115.

136 Thorner, 'Protestantism', 27–32.

137 Heller, *Labour*, 118.

138 Glick, 'Moriscos y marranos', 115–120, 124–127, Davids, *Rise and decline*, 227–229.

was not as able as in the past to compensate for the crumbling of its economic strongholds in international markets by building new ones. Richard Rapp has suggested that this failure can partly be blamed on the Venetian government, which mistakenly kept tariffs and taxes too high and strictly held on to a policy of prescribing high quality standards for export products.[139] State regulations did not entirely determine technology-in-practice, however, as Molà's case study of the silk industry in the sixteenth century has shown.[140] The decline in technological creativity may also have had to do with the progressive loss of command of markets for raw materials that would have permitted the Venetian economy to develop new specialties. In the seventeenth century, Venice lost a vital asset for technological leaders: a dominant position in a far-flung trading network.[141]

In the case of Spain, the decisive factor was not a loss of markets but the failure of the government to lend continued support to innovation in technology. The outburst of technological creativity in the sixteenth century largely took place under the aegis of state agencies and municipalities. Both central and local government had provided the principal incentives and favourable settings for technical innovation. Aside from municipal governments, the key institutions that sustained demand for technical innovation were agencies of the central state, such as the Royal Court, the *Casa de la Contratación*, the Council of the Indies, the Mint, the Army and the administration of mines. This was the audience for which prolific inventors like Pedro Juan de Laspanosa, Juan de Herrera and Jerónimo de Ayanz exerted all their powers of ingenuity.[142] But it was precisely the contraction of government demand in the early seventeenth century, due to the financial quandary of the Spanish state, that discouraged the further pursuit of technical innovations.[143] If the Spanish were 'underachievers' in technological creativity throughout most of the seventeenth and eighteenth centuries, this had more to do with a lack of support from the government than with their Catholic persuasions.

---

[139] Rapp, *Industry and economic decline*, esp. 140, 154, 159, 165.

[140] Molà, *Silk industry*, esp. 300–301.

[141] Davids, 'Shifts of technological leadership', 341–345.

[142] López Piñero, Ciencia, 96–111, 241–278, Goodman, *Power and Penury*, chapters 2 and 5, Goodman, 'Science, medicine, and technology', 12–24, García Tapia, *Téchnica y poder*, 13–20, 141–164, García Tapia and Carrillo Castillo, *Tecnología y imperio*, 38–43, 63–72, 85–91, 124–130.

[143] Galabert, 'Castile 1504–1808', 215–223, Davids, 'Dutch and Spanish global networks', Barrera-Osorio, *Experiencing nature*, 133–134; municipalities' financial difficulties from the 1620s also led to the reduction or cessation of subsidies for secondary education, see Kagan, *Students and society*, 45.

In France, by contrast, once the government had put a regulatory framework in place in the second half of the seventeenth century, it persisted in its encouragement of innovation. After c.1660, following tentative beginnings under the Valois kings in the mid-sixteenth century and during the reign of Henri IV, on the initiative of Louis XIV and his *contrôleur general des finances* Jean-Baptiste Colbert, the French state embarked on a continuous, systematic policy in support of inventors and innovations. Beside attracting large numbers of technical experts from abroad (notably from the Dutch Republic), the state stimulated technical innovation in the kingdom by offering exclusive privileges and all sorts of rewards for inventions, providing a rigorous procedure for assessing and validating their quality, supervised by the *Académie des Sciences*, and keeping up a large, permanent demand for technical innovations in the army, the navy and various state-controlled enterprises. The state offered an even more favourable setting for technological innovation in the later eighteenth century, when expectations regarding technology as an instrument for the reform of French economy and society ran ever higher. In the eighteenth century, inventors came to enjoy an exceedingly heroic, venerated position in French society – much more so, in fact, than inventive minds on the other side of the Channel.[144] The French state was, of course, not a monolithic mass. Inside the state apparatus, a variety of agencies had, in one way or another, something to do with sponsoring, or regulating, technological innovation. The *Conseil du commerce* and the *Bureau de commerce*, which both formed part of the *Contrôl general des finances*, were, along with the *Académie des Sciences*, most intensely involved in assessing and promoting the creation of new technologies, but other agencies, such as the *Maison du Roi* or the ministries of the army and the navy, were also increasingly concerned with the subject. Neither did institutions at the provincial and urban levels fail to become involved. Throughout the Old Regime, inventive activities in France remained much more oriented towards the state than the market. What inventors were primarily looking for was support from the government, not the exploitation of opportunities on domestic or international markets.[145]

In the seventeenth and eighteenth centuries, the countries in Europe where technical innovation proceeded at a higher rate and took place

---

144 Heller, *Labour*, chapter 4, Hilaire-Pérez, *L'invention*, 42, 51–65, 71–73, 143–146, 225–232, Alder, *Engineering the Revolution*, Minard, *Fortune du colbertisme*, Gillispie, *Science and polity in France*.

145 Hilaire-Pérez, *L'invention*, 36–38, 53–55, 71–82, 181, 319.

across a broader spectrum than in France were the Dutch Republic and England, successively. Both in the Netherlands and in England, market forces governed technical innovation to a greater extent than in France. Neither of these countries saw as massive an involvement of the state in technological innovation as had France during the later seventeenth and eighteenth centuries. This different balance between market and non-market mechanisms may better explain the variations in technological creativity between these countries than differences between 'Catholic' and 'Protestant' values, practices or institutions. Among the frontrunners in technology, however, it was not the Dutch Republic, but England, that first achieved a revolutionary breakthrough, which led to a fundamental shift in the structure of the economy at large.

What distinguished England from the Netherlands was *not* the presence of a more favourable set of institutions for technological creativity – all the institutions that favoured creativity in England existed in the Dutch Republic, too, and at a much earlier date[146] – but the fact that earlier than the Netherlands, England saw the rise of a public culture that offered a friendly setting for the spread of scientific, i.e. 'mechanical', Newtonian ways of thinking to engineers and entrepreneurs in industry. Thus the Scientific Revolution, in a sense, generated the Industrial Revolution through a more or less market-type process of supply and demand for new, 'useful' knowledge.[147] Joel Mokyr has called this link between the Scientific and Industrial Revolutions the 'Industrial Enlightenment' – a set of changes that transformed the two sets of useful knowledge, propositional and prescriptive knowledge, and the relation between them.[148] The Industrial Enlightenment thus influenced technical innovation, because it affected the resources from which new things can be fashioned, as well as the mental steps by which ideas for inventions could be turned into reality and the practical ways in which these could be shaped into techniques or artefacts that actually worked. In the Dutch Republic, by contrast, the relations between 'propositional' and 'prescriptive knowledge' in the field of industry developed to a lesser extent and at a slower pace than in Britain. The bottleneck was not a lack of expansion of propositional knowledge, nor

[146] Mokyr, *Enlightened economy*, 106–107, cf. Davids, *Rise and decline*, chapters 6 and 7 on the institutional environment in the Dutch Republic.

[147] See e.g. Inkster, *Scientific culture*, Stewart, *Rise of public science*, Golinski, *Science as public culture*, Jacob, *Scientific culture*, Goldstone, 'Efflorescences', Mokyr, *Enlightened economy*, 40–62, 122.

[148] Mokyr, *Gifts of Athena*, 35, Mokyr, *Enlightened economy*, 30–40.

a lack of transmission between the two sets of useful knowledge, but an absence of growth of 'useful' knowledge for industrial purposes.[149]

The forging of this link between science and industry in Britain during the eighteenth century was a dual process. On the one hand, people who claimed to be experts in science made determined efforts to spread their wisdom to a wider audience. Their aim was to win science its place in society by turning it into a 'public' enterprise. Bringing science into public view by writing tracts or text-books, by holding lectures in coffeehouses and societies, by making ingenious instruments or by performing experiments that allowed anyone who wished to attend to gain insights into the workings of nature, these practitioners of 'natural philosophy' hoped to establish some legitimacy for their way of pursuing knowledge which, as a perceived threat to the prevailing religious and political order, it did not yet possess in the early years of the Royal Society. While at first, scientific enterprise was mainly defended as a prop for religion, from the early eighteenth century onwards, it was increasingly vindicated in terms of usefulness as it was supposedly proven in practice. The achievements of Isaac Newton were the principal basis for these claims.[150] Usefulness remained a constant theme in the vindication of 'natural philosophy' and science in Britain from the seventeenth to the nineteenth century.[151] On the other hand, champions of science were able to reach a wider audience precisely because important groups in society were fully prepared to support this pursuit of scientific research for practical ends. In England, it was among the landed gentry that natural philosophers in the early eighteenth century first found a source of support, later joined by merchants, manufacturers and the urban middle classes at large.[152] Whether science actually yielded as many practical benefits as its champions claimed mattered less than the firm belief that it did.[153]

While efforts to legitimise science were also undertaken in the Dutch Republic long before the middle of the eighteenth century, the utilitarian argument did not come to the fore until after 1750. For a long time, science was almost exclusively justified with the argument that inquiry

---

149 Davids, *Rise and decline*, chapter 7.

150 Stewart, *Rise of public science*, part III, Golinski, *Science as public culture*, chapters 1 and 2, Schaffer, 'Natural philosophy'.

151 Gaukroger, *Emergence*, 39–40.

152 Stewart, *Rise of public science*, part III, Golinski, *Science as public culture*, chapters 1 and 2, Schaffer, 'Natural philosophy'.

153 Cf. Gaukroger, *Emergence*, 41, who surprisingly does not discuss studies by Stewart, Golinski, Inkster, Jacobs, Mokyr, Schaffer or Stewart quoted above.

into nature provided perfect proof for the existence and omnipotence of God.[154] Producing 'useful' knowledge for industry did not yet form part of the repertoire of arguments to defend science. Why did the development of scientific thinking in the Netherlands differ from that in Britain? Part of the explanation is that a source of support for science that had proved to be critical in England during the early decades of the eighteenth century was only marginally present in the Dutch Republic. While science in the Netherlands, like in Britain, had an increasingly wide audience among the urban middle classes from the 1730s onwards, for much of the eighteenth century in this country, the figure of the 'improving' landlord, who called on the services of chemical practitioners or natural philosophers in all sorts of ventures to enhance the value of his estate, only played a modest, background role. This 'demand' factor was absent. The absence of the 'improving' landlord in the Netherlands was not merely a consequence of the different ways in which ecclesiastical property seized during the Reformation in the sixteenth century was disposed of. While the dissolution of the monasteries in England since the 1530s was doubtless a major factor in the expansion of the size of landed estates held by the gentry, as Philip Gorski in his revised version of the Weber thesis suggests, it did not create large landholdings as such.[155] In the Netherlands, too, part of the land seized from the Church in the 1570s was sold to private buyers, but the shift in ownership did not lead to the consolidation of large, landed estates.[156] The average size of landed property in the Netherlands remained much smaller than in England. The difference in the structure of landholding in England and the Netherlands, which had already emerged in the High Middle Ages,[157] was accentuated rather than created by the Reformation.

Variations in religious contexts *did* influence the ways in which natural philosophy was defended, but this influence operated at a deeper level. Charles Taylor has pointed out that one of ways in which we can understand secularization is as the 'shift from a society where belief in God is unchallenged and ... unproblematic to one in which it is understood

[154] Davids, *Rise and decline*, 523.

[155] Gorski, 'The Little Divergence', 180–181, Hoskins, *Age of plunder*, 135–138, Overton, *Agricultural revolution*, 168–169.

[156] Van Beeck Calkoen, *Onderzoek naar den rechtstoestand*, 249–254, Maris, *Reformatie der geestelijke en kerkelijke goederen*, 214–245, Van Apeldoorn, *Kerkelijke goederen*, vol. II, 422–423, Roelfsema, *Klooster- en proosdijgoederen*, 76–100.

[157] Hoppenbrouwers and Van Zanden, 'Restyling the transition'.

to be one option among others'.[158] This great transformation started in Europe in the early modern period. How did the 'alternatives become thinkable'? An important part of the answer, in Taylor's view, is the coming of the Protestant Reformation, which launched a massive attack on 'the enchanted world' in which most people lived in about 1500, and which, through the drive to re-order society in its church structure and the personal life of its members, prepared the ground for the rise of self-sufficient, 'exclusive' humanism, which is one of the principal alternatives to belief in God today. The new religious and moral order, as envisaged by the reformers, 'had no place for the ambivalent complementarities of the older enchanted world.... it was coherent, uncompromising, all of a piece'.[159] It was the ultimate consequence, in a sense, of the disputative approach widely adopted by scholars in Europe from the High Middle Ages. From the early sixteenth century, the synthesis between Christian theology and Aristotelian natural philosophy with the aid of metaphysics, which had been wrought by scholastic philosophers in the thirteenth century, also increasingly came under attack from various quarters.[160] In a culture that was still overwhelmingly Christian, this implied that another strategy had to be devised to defend the legitimacy of natural philosophy in a way that was compatible with Christian beliefs. For Protestants, this meant, above all, compatible with revelation as found in Scripture.

From the mid-seventeenth century onwards, the solution that found increasing support among intellectual elites was that of physico-theology. Physico-theology was based on the assumption that inquiries into nature would lead to a fuller understanding of God's design of the world and his intentions for his creation. 'The idea that natural philosophy is a means of seeking evidence of God's activity in nature [became] widespread in the 1680s and 1690s, particularly in England', Stephen Gaukroger observed.[161] Physico-theology found a large group of adherents in the Netherlands, too. But there was a difference in emphasis: physico-theological writings in the Netherlands, especially *Het regt gebruik der wereltbeschouwingen* published by Bernard Nieuwentijt in 1715, bore a much more orthodox stamp than similar works produced in Britain. In the English translation,

158 Taylor, *Secular age*, 3.
159 Taylor, *Secular age*, 19, 77, 80, 146.
160 Gaukroger, *Emergence*, 129–153, 227, 506.
161 Gaukroger, *Emergence*, 493–505.

Nieuwentijt's work was even shorn of its very references to traditional Christian doctrine.[162]

Why was that? The vindication of natural philosophy in the Dutch Republic had to take place in a somewhat different atmosphere from that in England. Church and State in the Dutch Republic were not as closely intertwined as in England in the later seventeenth and eighteenth centuries. In the Netherlands, open or tacit sympathy for the cause of natural philosophy in political circles did not correspond as neatly with support from the sphere of the Church as in England. In England, the ascendancy of the 'latitudinarian' or 'low church' tendency within the Church of England, which generally took a more friendly view of science than the 'high church' party, corresponded with the growing power of the equally sympathetic Whigs in the sphere of politics.[163] In the Dutch Republic, by contrast, the favourable attitude of political elites towards natural philosophy was not mirrored by the state of affairs in the Dutch Reformed church. In the Reformed Church, it was the orthodox tendency that reigned supreme in the later seventeenth century and first half of the eighteenth century.[164]

Moreover, moderate Calvinist believers in the Netherlands faced another formidable opponent much earlier than 'latitudinarians' in England: a group of radical philosophers who questioned the foundations of religion itself. In the second half of the seventeenth century, the Dutch Republic was the birthplace of what has become known as the 'Radical Enlightenment'. After Cartesianism had made its breakthrough in the Dutch academic world in about 1650, a number of thinkers, notably Baruch de Spinoza, carried the new philosophy to extremes, by attacking the divine authority of the Bible or even doubting the very existence of God.[165] After that, anyone who wished to defend a new approach to science had to cover his flank against the radical philosophers, who put a higher value on independence of mind than any form of traditional belief, in order to be better able to counter criticism from conservative

---

[162] Vermij, *Secularisering*, chapters 3 and 6. Zuidervaart, *Van 'konstgenooten'*, 70–71, 360–362, Van Berkel, *Citaten*, 8, 22, Jorink, *Het Boeck der Natuere*. On Nieuwentijt and the rise of Dutch physico-theology in general, see especially Bots, *Tussen Descartes en Darwin*.

[163] Hunter, *Science and society*, 135, 188, Hunter, *Science and the shape of orthodoxy*, 110, 118–119.

[164] Vermij, *Secularisering*, 88–136, Van den Berg, 'Orthodoxy', Roorda, 'Contrasting', 140, Mijnhardt, 'Dutch Enlightenment', 214–215.

[165] Klever, *Mannen rond Spinoza*, Israel, *Radical Enlightenment*, Van Bunge, 'Introduction', Van der Wall, 'Religious context', 39–43, Mijnhardt, 'Construction of silence', 231–232, 256–259.

theologians, who placed godliness above 'vain learning'.[166] Apologetics for the sake of religion became a necessary condition for the justification of science. 'By attacking Spinoza, his radical Cartesian colleagues and their popularizers', Wijnand Mijnhardt has remarked, the

> moderate philosophers and theologians expected to blunt the attack of the orthodox ... on themselves. They also hoped to show that the tenets of the new philosophy and the new visions of science, politics and theology that were based upon it, did not necessarily lead to atheism; and finally they wanted to propagate themselves as the defenders of the religious and moral order.[167]

In this context, vindicating science in terms of practical usefulness was *less* helpful in securing a 'niche' for science than using arguments stressing its 'religious correctness'. In so far as science in the Netherlands in the late seventeenth and first half of the eighteenth century had to be defended, it was threatened, on the one hand, by the powerful orthodox tendency within the Reformed church, and on the other hand, by radical Cartesians and Spinozists. Champions of science only began stress its practical usefulness after the middle of the eighteenth century, when orthodoxy no longer reigned supreme within the Reformed church, the radicalism of the Dutch Enlightenment had been tamed and the political context was changing as well. As the clamour for economic reform grew louder, natural philosophy could more easily be presented as an activity with manifest practical uses. From the later eighteenth century, science was justified as an eminently patriotic enterprise.[168] But by then, the industrial breakthrough in Britain had already begun.

Conclusion

To what extent can differences in technical innovation between China and Europe and within Europe be explained by variations in religious contexts? This is the key question that this chapter has sought to answer. Before the sixteenth century, developments in China and Europe ran parallel in more than one respect. In this period, examples of religious

---

166 Van der Wall, 'Religious context', 52–53.

167 Mijnhardt, 'Construction of silence', 231–232.

168 Mijnhardt, *Tot heil van 't menschdom*, 88–89, 106–112, idem, 'Dutch Enlightenment', 216–222, idem, 'Construction of silence', 234–238. Davids, 'From De la Court to Vreede', 270–278, Theunissen, *Nut en nog eens nut*, 13–16.

institutions stimulating innovation in the building industry, hydraulic technology and other branches of technology were not hard to find in either region. After 1500, however, the impact of religion on the creation of inventions in China and Europe increasingly diverged. In China, to a greater extent than in Europe, technical innovation took place *outside* the context of religious institutions. A distinctive feature of China was the overriding importance of the state as a force in technological change. Incentives that allowed individuals to reap personal rewards for inventive activities were not introduced. Market demand for the services of technical experts like architects or engineers hardly developed.

In Europe, secular actors and institutions became progressively more important in technological innovation from the Late Middle Ages onwards. The expansion of the market economy, the rise of urban noble courts and the growth of patronage by secular governments opened up a range of opportunities for a creative class of technical practitioners and artisans. New incentives for innovation were introduced, such as patents, which consisted of a mixture of individualized rewards and provisions to benefit the community at large. But neither the settings nor the incentives for innovation became purely secular. Settings and incentives formed part of a composite mixture that included both religious and secular elements. A multitude of actors and institutions was involved, ranging from popes, bishops and religious orders via princes, dukes and noblemen to cities, merchants and craft guilds. After 1500, religious institutions in Europe continued to make significant contributions to the creation of technical inventions. The Catholic Church still lent massive support to the arts and the building industry, especially in papal Rome, which offered an exceedingly friendly environment for innovations.

China and Europe also began to differ with respect to other aspects of innovation. While in China after the fourteenth century, techniques of graphic representation did not improve to the extent that illustrations could give a realistic image of machinery of any complexity, the uses and styles of machine drawings in Europe kept evolving for a very long time. One of the uses of such drawings was to serve as a means to 'think' on paper. 'Thinking' on paper sometimes even took the form of theoretical reflection on the properties of machines with the help of geometrical analysis. Styles of machine drawings in Europe evolved from 'logical schemas built of icons' to 'depictions of appearance' – from pictures which gave a general idea of the main parts and the general arrangement of a mechanical device to drawings which showed what machines actually looked like.

Other differences emerged in the content and interrelations between the spheres of 'prescriptive' and 'propositional' knowledge. Scholarship and technology in China hardly became connected at an analytical level. Chinese learning concentrated on documenting, cataloguing and finding correlations. Even if Chinese scholars *did* show some interest in uncovering regularities and general patterns, they did not normally think in terms of 'laws of nature'. Engineers or craftsmen, for their part, did not put general reflections about their practice on record. Thus, 'propositional' knowledge only grew in a partial way and the spheres of 'prescriptive' and 'propositional' knowledge did not become strongly interlinked, which constrained innovation in ways other than through 'learning by doing' or 'learning by using'. In late medieval and early modern Europe, by contrast, it was not only 'the observation, classification, measurement, and cataloguing of natural phenomena' which formed an expanded body of propositional knowledge, but also 'the establishment of regularities, principles and "natural laws" that govern these phenomena and allow us to make sense of them'. And technical experts, with the help of 'propositional knowledge' of mechanics, also began to include theoretical reflections on properties of machines in their drawings more often. Thus in contrast with China, prescriptive and propositional knowledge in Europe gradually became more intertwined.

Like the changes in settings and incentives, these cognitive changes in Europe were to no small extent made possible by religious factors. It was religious forces, starting with the 'Papal Revolution' in the High Middle Ages, that got the process going, whereby the 'prescriptive' and 'propositional' spheres of 'useful' knowledge gradually became more closely interlinked than was the case in contemporary China. Religious institutions drove this process for a very long time, thanks to their vital contributions to the formation of human capital, via universities and other institutions for elementary, technical and vocational schooling, and thanks to their important role in the long-distance circulation of technical knowledge.

Technological creativity differed between regions within Europe, too. Between the High Middle Ages and the Industrial Revolution, Europe witnessed the rise and decline of a sequence of technological leaders. 'Technological leadership', which implied that a given country, region, town or cluster of towns played an initiating role in the development of new technologies in a wide variety of fields for a prolonged period of time, shifted from one area of the Continent to another. The direction of these shifts roughly ran from southeast to northwest: leadership moved from Northern Italy via Southern Germany and Flanders to the Dutch Republic and England.

At times, centres of technological creativity also existed outside this central belt, for instance in France or in the Iberian Peninsula, but these places did not play an initiating role in the development of new technologies in such a wide variety of fields for such a lengthy period of time as the regions that functioned as technological leaders.

The movement of the centres of technological leadership towards the North Sea area ran, geographically, more or less parallel with the 'Little Divergence': the growing economic divergence between ascendant areas in north-western Europe, notably the Netherlands and England, and 'less dynamic' regions in other parts of the continent, such as France, Spain and the Italian peninsula. Many sociologists, economists and economic historians will be tempted, of course, to seek the explanation for this phenomenon in some version of the Weberian argument about the relevance of the 'Protestant ethic': the Reformation was good for technical innovation and economic prosperity, the Counter-Reformation was bad. But reality was not that simple. In human capital formation, in the Late Middle Ages and early modern period, 'southern' and 'northern', or 'Protestant' and 'Catholic', regions of Europe were not fundamentally different. Neither were the Protestant and Catholic Reformations, on balance, an obstructing factor for the circulation of technical knowledge. To a large extent, different regions in Europe had shared experiences of human capital formation and the circulation of knowledge. The rise of technological leaders did not depend on the presence of a specific set of religious beliefs or values, but on a combination of particular favourable economic, political and social conditions. The slowdown, or lag, in technical innovation in countries like the Venetian Republic, Spain and France in the seventeenth and eighteenth centuries turns out, on closer inspection, not to be related to the 'retarding' force of the Counter-Reformation, but to developments in trade networks, changes in state finances and the nature of balances between state regulation and market mechanisms.

In fact, variations in religious contexts were more relevant for the emergence of differences in technological creativity between countries where Protestantism reigned supreme: in the Dutch Republic and England in the eighteenth century. The different relationship between Church and State, the prolonged dominance of the orthodox tendency in the Reformed Church and the early rise of the Radical Enlightenment may help to explain why natural philosophy in the Netherlands experienced its utilitarian turn much later than Britain. And that was one of the reasons why the advance of knowledge in the Dutch Republic did not produce an 'Industrial Enlightenment' such as that which led to the great breakthrough in technological creativity in Britain in the late eighteenth century.

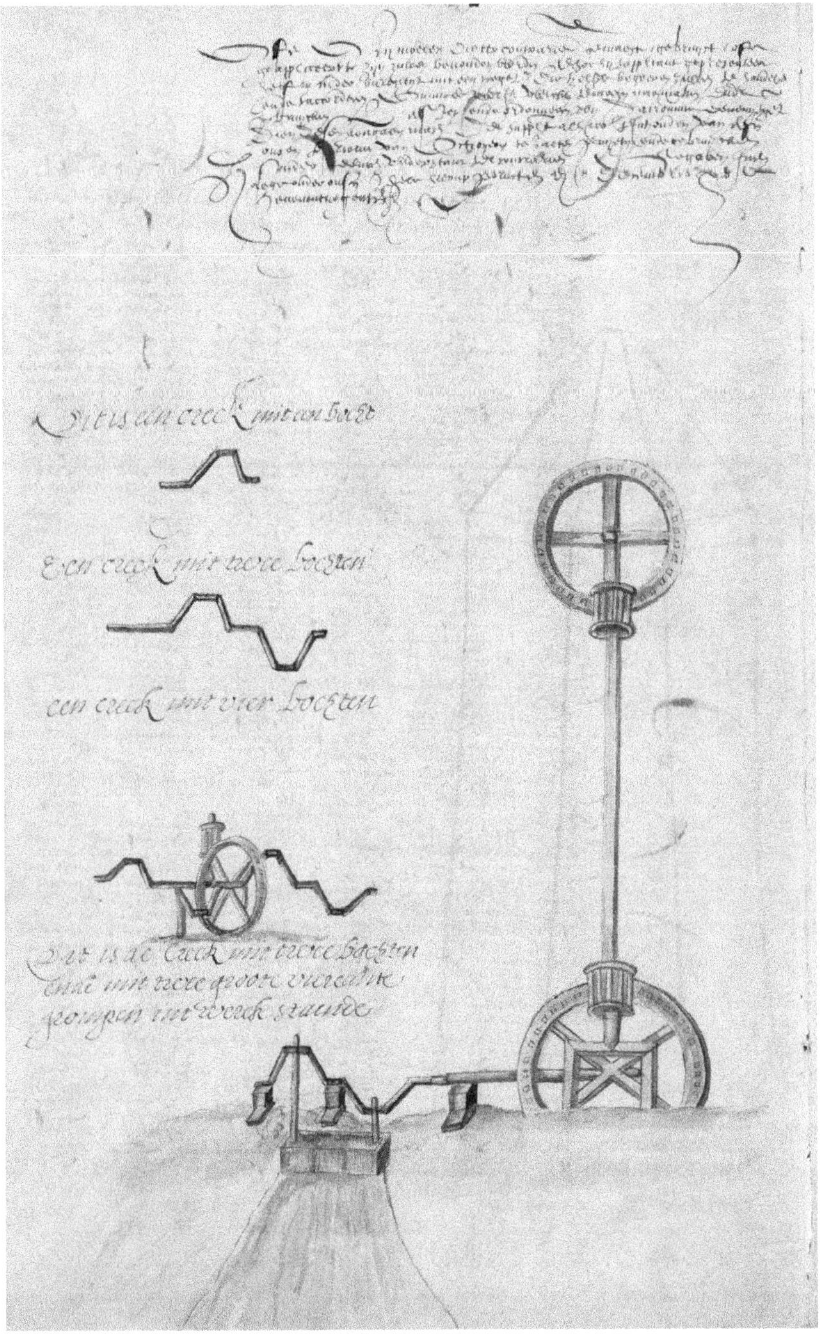

Figure 4. Drawing of a crankshaft for a wind-powered saw-mill in a patent application submitted to the States of Holland (United Provinces) by Cornelis Cornelisz. van Uitgeest, 1597. Courtesy Nationaal Archief, The Hague, The Netherlands.

CONCLUSION

Many ghosts haunt the history of technology in Europe and China before 1800: the ghosts of Su Song, Zheng He, George Macartney, Lynn White and Max Weber. The ghosts of Su Song, Zheng He and George Macartney seem to tell us that for a long time, the Chinese were able to pull off feats that Europeans could only dream of, but that for some reason or another, they chose not to pursue further technological advancement, even though they did not lose their technological edge until after 1800. Another reading of this message is that the Chinese had reached a technological dead end by 1500, and unwisely ignored opportunities to learn from the West in the centuries thereafter. The ghost of Lynn White loudly proclaims that the West was technologically superior before the end of the Middle Ages, but that the ruthless exploitation of the physical environment by technological means, inspired by the attitudes of Latin Christendom towards nature, unfortunately also became the root cause of the ecological predicament in which the world found itself at the end of the twentieth century. Christianity should therefore take a huge part of the blame for our present ecological woes. And the ghost of Max Weber whispers that a special brand of Christianity, namely ascetic Protestantism, was much more favourable to economic prosperity and technological advance than the traditional variant, Roman Catholicism. Protestantism represented the forces of progress, while Catholicism was the incarnation of backwardness.

This book has aimed to take a fresh look at the long-term evolution of technology in China and Europe. More specifically, it has sought to compare the development of technology in China and Europe in relation to variations in religious contexts. Although technology and religion routinely come up in narratives about differences in economic performance between China and Europe (the 'Great Divergence') and between north-west and south-east Europe (the 'Little Divergence'), which became manifest at some time in the early modern period, the connections between the two have not been examined in a systematic, comparative way. Some scholars emphatically claim that the European 'invention of invention' was rooted in a distinctive set of religious values,[1] while others,

[1] Cf. Landes, *Wealth and poverty*, 58–59.

equally emphatically, claim that before 1800, 'no major cultural distinctions between China and Europe', not even in 'religious doctrines . . . can be detected that had unequivocally differential . . . effects on economic growth and technological change'.[2] But are such claims supported by the historical evidence? Neither group has tried to find out.

Making systematic comparisons is exactly what this book has set out to do. The focus of the study has not been on proximate causes of technological change, such as relative factor prices, but on underlying factors that affect the supply of 'useful' knowledge. It has concentrated, more specifically, on the extent to which these underlying factors have been influenced by variations in religious contexts. By taking this particular angle on the problem, this book makes an innovative contribution to the debate on the 'Great' as well on the 'Little Divergence'. The key question addressed in this book is, to what extent can disparities in the evolution of technical knowledge between Europe and China and between regions within Europe from about 700 to 1800 be explained by differences in religious context? The answer has a negative side and a positive side.

The negative part is that the familiar ghosts turn out to have lost all their credibility. None of the traditional interpretations of the stories of Su Song, Zheng He and Macartney withstand closer scrutiny. Su Song's device remained unique in its kind, yet clock-making in China continued to evolve further throughout Yuan and early Ming times. Zheng He's expeditions were more a repeated show of force in well-known areas than a sequence of true voyages of discovery. The fate of Macartney's embassy does not tell us anything about Chinese or British technological superiority at the end of the eighteenth century. The reality of technological change was much more complex than conventional readings of such iconic events suggest. China was not the only cradle of invention in the world before 1500, but neither was early modern Europe more technologically advanced than China in every respect. Between c.700 and 1800, both China and Europe went through long periods of incremental technological change, alternated with outbursts of accelerated creativity. Eventually, a revolutionary technological breakthrough was achieved in Europe and not in China, but this should not lead us to exaggerate differences in the general *level* or *rate* of technological change in Europe and China before the end of the eighteenth century.

---

2 Perdue, 'How different was China?', 311–312.

Although the general level and rate of technological change in China and Europe before the end of the eighteenth century did not vary greatly, there were nevertheless some salient differences in the *nature* of technological change. Technological evolution in China differed from developments in Europe in the sense that particular techniques or artefacts that were initially common to both regions of Eurasia eventually evolved to a lesser extent in China than in Europe. The potential of specific innovations was not always realized as fully in China as in Europe, even if they *did* come into widespread use. With respect to 'mechanical' technologies, Europe followed a more *consistent* trajectory than China, but the economically most advanced parts of Europe, such as the Netherlands, also continued to advance in 'skill-oriented' technologies throughout the early modern period. Contrary to Mark Elvin's thesis that Late Imperial China suffered from 'a relative technological standstill' due to the so-called 'high-level equilibrium trap', Qing China still saw significant, incremental technological change in the agricultural sector. Technological change in early modern Europe, by contrast, had a wider scope: it took place both in the agricultural sector and in urban industries, as well as in other economic sectors, such as the shipping industry. Neither China nor Europe after 1000 went through an age of technological 'regression', in the sense of a lengthy period during which important knowledge was lost. In both regions, there was evidently a kind of ratchet effect at work. 'Regression' in technological development, however, in the sense of a prolonged, large-scale shift from a technology of some complexity to a more simple technique, was not an uncommon occurrence in China after the Song.

The ghosts of Lynn White and Max Weber can be laid to rest as well. Contrary to what Lynn White and others have claimed, variations in visions of the uses of nature between religious traditions between c.700 and 1800 were too weak or too inconsistent to offer a plausible explanation for differences in technological development. Latin Christendom did not invariably condone the ruthless exploitation of nature. In fact, Scripture allows for a variety of attitudes towards the natural world, ranging from exploitation to stewardship. There is no evidence that from the Middle Ages onwards, Christians in Europe generally preferred the former to the latter. Neither did Latin Christendom have a monopoly on images of domination of nature. Ideas and practices conducive to the exploitation of nature could be found before the advent of Christianity and also existed in parts of the world where entirely different religious traditions reigned supreme, such as the Chinese Empire.

The Weberian approach has been found wanting, too. Time and again, conjectures based on Weber's work have failed to offer a convincing explanation for differences in technological achievement inside Europe. Whether one examines visions on the uses of nature, human capital formation, the circulation of technical knowledge or technical innovation, no clear relationship with 'ascetic Protestantism' emerges. Calvinism and Puritanism did not stimulate the exploitation of nature any more than Catholicism did, they were not better than Catholicism in promoting literacy or providing facilities for formal schooling, they did not lead to a more rapid or more extensive circulation of knowledge and neither did they offer a more congenial environment for technological creativity. At times, differences within Protestantism seem to have mattered more than differences between 'ascetic Protestantism' and Catholicism. The different relationship between Church and State in England and the Dutch Republic in the seventeenth and eighteenth century, plus the prolonged dominance of the orthodox tendency in the Reformed Church and the early rise of the Radical Enlightenment in the Netherlands, may help to explain why the utilitarian turn in natural philosophy occurred much later in the Netherlands than on the other side of the North Sea. In general, Weber-inspired stereotypes obfuscate matters rather than clarify them. In this case, the Weberian approach leads us nowhere.

Explaining variations in the evolution of technical knowledge by looking at the impact of religious ideas, attitudes and values, as the 'schools' of White and Weber attempt to do, has thus proved to be a singularly sterile exercise. An alternative strategy turns out to be much more productive: taking not the sphere of ideas, attitudes and values as a primary focus, but concentrating on institutions, patterns of communications and movements of people as the chief points of religious impact. This, in turn, is the positive side of the answer to the key question addressed in this book.

The supply of technical knowledge in China and Europe over time, and the role played by religious contexts in this supply of knowledge, was studied from three angles: the formation of human capital, the circulation of knowledge and technical innovation. Human capital and the circulation of knowledge and skills, discussed in Chapters 2 and 3, lie at the beginning and the end of the 'loop' of technological change. Human capital is both an ingredient and a product of technological change – and the same holds for the circulation of knowledge and skills. The middle section of the loop, technical innovation, formed the subject of Chapter 4.

The essentials of the story that emerges from the long-term comparisons are as follows. Both in China and in Europe, human capital formation and the circulation of knowledge after c.700 developed to such an extent that they provided effective steps in the evolution of technology, in the sense that they prevented important knowledge from ever being completely lost. Technological 'regression' in the hard sense of the word did not occur. Both in China and in Europe, informal learning throughout the pre-industrial period remained of crucial importance for abilities that rested, to a relatively large extent, on 'tacit' knowledge, such as skills in extracting, shaping or handling materials and material objects. The pattern of human capital formation, circulation of knowledge and technical innovation in China and Europe nevertheless began to diverge from the fifteenth century onwards.

Although China from the Song saw significant growth in its system of formal education, accompanied by at least some growth in literacy, until the middle of the nineteenth century, schools with a technical or vocational emphasis remained few and far between. However much the state was prepared to encourage the expansion of schooling, it hardly concerned itself with facilities for specialist, technical training. Up to the nineteenth century, technical education in China largely took place via informal learning and essentially remained a matter of family tradition.[3] Moreover, the institutional setting for the regular, long-distance mobility of artisans created by the early Ming disappeared in the sixteenth century, even though the examination system for state officials remained fully operational. The output of printed books, in addition to the continued production of texts in manuscript form, increased markedly from the late fifteenth century onwards, but the accessibility of these materials in print and script, including technical writings, did not grow accordingly. Fixed sites for the collection and storage of technical knowledge in the form of objects, such as botanical gardens, curiosity cabinets, model collections or museums, were exceedingly rare. As far as technological literature is concerned, the output of books about agriculture vastly exceeded the number of writings on machines and mechanical devices. Furthermore, even though the art of making technical drawings in China had reached a high level of sophistication by the early fourteenth century, techniques of graphic representation did not improve thereafter to the extent that illustrations could give a realistic image of machinery of any complexity. Chinese technicians did not start to use drawings to 'think on paper'.

[3] Lee, *Education*, 512–541.

Scholarship and technology in China hardly became connected at an analytical level. Chinese learning concentrated on documenting, cataloguing and finding correlations. Even if Chinese scholars *did* show some interest in uncovering regularities and general patterns, they did not think in terms of 'laws of nature'. Engineers or craftsmen, for their part, did not put general reflections on their practices on record. Thus, 'propositional' knowledge grew only in a partial way and the spheres of 'prescriptive' and 'propositional' knowledge did not become strongly interlinked, which constrained innovation in ways other than through learning by doing or learning by using.

In Europe, by contrast, the expansion of formal education from the Late Middle Ages also included the growth of vocational and technical schooling. Long-distance movements of skilled people increased in many parts of Europe through the spread of the practice of tramping of journeymen. As in China, the output of printed books, including technical writings, vastly increased from the late fifteenth century, but, in contrast with China, libraries became more numerous and more easily accessible to various groups of people. In Europe, artefacts, especially models, also gradually became more important for the circulation of knowledge. The network of fixed sites where technical information was collected and stored assumed a different structure from that which emerged in China. Aside from libraries, botanical gardens, curiosity cabinets, model collections and museums became integral parts of the network as well. Amongst the expanding body of technological literature were a growing number of books on machines and mechanical devices that recorded both technology-in-use and projected images of technology that might be developed in future. In contrast with China, European technical illustrations steadily became more realist, detailed and precise. Styles of machine drawing in Europe evolved from pictures that gave a general idea of the main parts and the general arrangement of a mechanical device, to drawings that quite accurately showed what machines actually looked like. And the uses of drawings kept evolving, too. One of the new uses of technical drawing was to serve as a means to 'think' on paper. 'Thinking' on paper sometimes even took the form of theoretical reflection on the properties of machines, with the help of geometrical analysis.

Moreover, in late medieval and early modern Europe it was not only 'the observation, classification, measurement, and cataloguing of natural phenomena' which formed an expanded body of propositional knowledge, but also 'the establishment of regularities, principles and "natural laws" that govern these phenomena and allow us to make sense of them'.

Technical experts, with the help of 'propositional knowledge' of mechanics, also began to include theoretical reflections on properties of machines in their drawings more often. Thus in contrast with China, prescriptive and propositional knowledge in Europe gradually became more interlinked, which expanded the base for the creation of new technologies.

Variations in religious context *did* matter for this divergence between China and Europe, but in ways that were different from those imagined by the schools of Lynn White and Max Weber. In this book, religion has been regarded as a multi-faceted phenomenon, which finds expression in people's beliefs, practices and experiences, in institutions and media, as well as in rituals, images and doctrines. Following Steven Bruce, it has been conceived as an ensemble of 'beliefs, actions, and institutions, which assume the existence of supernatural entities with powers of action, or impersonal powers or processes possessed of moral purposes'. Religion in this sense has particularly influenced the evolution of technology through institutions, patterns of communications and movements of people. The root of the disparities between China and Europe, in particular from the fifteenth century onwards, lay in the varying relations between governments, markets and religious institutions. In China from the Song onwards, the balance between these three forces in the field of technology tilted heavily towards the central government. The reduction of the weight of religious institutions was only partly compensated by the expanding share of the role played by markets. In Europe, the weight of governments and markets since the High Middle Ages in human capital formation, circulation of knowledge and the creation of new technologies had substantially increased, but religious organizations, in particular those of the Catholic Church, continued to have a strong impact in all three areas as well.

Whereas religious institutions in China from the Song onwards were increasingly relegated to a subordinate position in informal and formal learning, monasteries and church building sites in Europe, especially between c.1100 and 1500, but also thereafter to some degree, offered craftsmen plenty of opportunities for informal and formal learning. Before 1500, religious institutions in Europe did not only make a much larger contribution to the supply of formal education than state governments, but they continued, or once again began, to be significant suppliers of educational facilities even after secular governments and commercial entrepreneurs entered the field. From the sixteenth century onwards, religious institutions, especially those in Catholic regions, made a remarkable comeback in elementary, secondary and higher education. In many places in Europe, they were at the forefront of the development of vocational and technical training. The reason why religious institutions in Europe were able to act more independently than in China, even though the material resources of ecclesiastical institutions in Latin Christendom before 1500 may have been no larger than those of Buddhist or Daoist institutions in Tang, Song or Yuan China, was that their organizational capabilities increased substantially from the High Middle Ages onwards. Between about 1050 and 1250, the Church in Latin Christendom, through the twin movements of the 'Papal Revolution' and the emergence of formally linked networks of monastic orders, successfully achieved a high degree of internal cohesion and a large measure of independence with regard to secular authorities, which, in Catholic regions at least, it largely managed to retain during much of the early modern period.

Religious institutions, rituals and practices in Europe, moreover, did more to promote the circulation of technical knowledge than similar phenomena in China. While the growth of pilgrimage and the creation of regular communication between formally 'networked' monastic houses in Europe in all likelihood began to stimulate long-distance flows of technical knowledge in Europe during the High and Late Middle Ages, such patterns of religious travelling did not develop to the same extent in China. From the sixteenth century, the Catholic Reformation gave a further boost to the circulation of technical knowledge in important ways. Missionary fervour spurred long-distance travel by Europeans, which contributed to a further expansion and acceleration of the circulation of technical knowledge and skills. 'Fixed places' for the collection and storage of technical information, such as libraries, gardens, curiosity cabinets, model collections and museums, which arose in many places in Europe in the early modern period, were partly created by the very organizations, such as the Society of Jesus, that battled in the forefront of the educational and missionary campaigns of the Catholic Reformation.

In the creation of new technologies, too, religious variables had a much greater impact in Europe than in China. Before 1500, examples of religious institutions stimulating innovation in technology can be found both in China and in Europe. After 1500, however, to a greater extent than in Europe, technical innovation in China took place *outside* the context of religious institutions. In Europe, the expansion of markets, the rise of urban noble courts and the growth of patronage on the part of secular governments from the Late Middle Ages admittedly opened up a range of opportunities for a creative class of technical practitioners and artisans who pioneered the changes in writing, drawing and thinking described above. New incentives for innovation were introduced, such as patents, which consisted of a mixture of individualized rewards and provisions to benefit the community at large. But neither the settings nor the incentives for technical innovation became purely secular. The settings and incentives that favoured technological creativity included both religious and secular elements. The Catholic Church still lent massive support to the arts and the building industry, especially in papal Rome, which provided a favourable atmosphere for innovations. Religious forces, too, starting with the 'Papal Revolution' in the High Middle Ages, triggered the long-term process whereby the 'prescriptive' and 'propositional' spheres of 'useful' knowledge gradually became more intertwined. Thanks to their seminal role in the formation of human capital and their significant contributions to the long-distance circulation of technical knowledge, religious organizations would drive this process in Europe for a long time to come.

This book started with a question about the extent to which disparities in the evolution of technical knowledge between Europe and China and between regions within Europe between about 700 and 1800 can be explained by differences in religious context. The answer can be summed up in a few words: the Little Divergence in economic performance within Europe owed little to religious forces, but religion mattered a great deal for the Great Divergence between China and Europe.

BIBLIOGRAPHY

Adas, Michael, *Machines as the measure of men. Science, technology, and ideologies as measure of Western dominance* (Ithaca 1989).
A'Hearn, Brian, Jörg Baten and Dorothee Crayen, 'Quantifying quantitative literacy. Age heaping and the history of human capital', *Journal of Economic History*, 69 (2009) 783–808.
Aiyar, Shekhar, Carl-Johan Dalgaard and Omer Moav, 'Technological progress and regress in pre-industrial times', *Journal of Economic Growth*, 13 (2008) 124–144.
Albrecht, Helmuth, 'Die Anfang des technischen Bildungssystems', in: Laetitia Böhm and Charlotte Schönbeck (eds.), *Technik und Bildung* (Düsseldorf 1989) 118–153.
al-Hassan, Ahmad Y. and Donald R. Hill, *Islamic technology. An illustrated history* (2nd edn., Cambridge 1992).
Alder, Ken, *Engineering the Revolution. Arms and Enlightenment in France 1763–1815* (Princeton 1997).
Allen, Robert C., 'Collective invention', *Journal of Economic Behavior and Organization*, 4 (1983) 1–24.
——, 'Economic structure and agricultural productivity in Europe, 1300–1800', *European Review of Economic History*, 3 (2000) 1–25.
——, *The British Industrial Revolution in global perspective* (Cambridge 2009).
Allmann, Joachim, *Der Wald in der frühen Neuzeit. Eine mentalitäts- und sozialgeschichtliche Untersuchung am Beispiel des Pfälzer Raumes 1500–1800* (Berlin 1989).
Allsen, Thomas T., *Culture and conquest in Mongol Eurasia* (Cambridge 2001).
Amery, Carl, *Das Ende der Vorsehung. Die gnadenlose Folgen des Christentums* (Hamburg 1972).
Amouretti, Marie-Claire, 'L'attelage dans l'Antiquité. Le prestige d'une erreur scientifique', *Annales ESC*, 46 (1991) 219–233.
Anthiaume, Albert, *Évolution et enseignement de la science nautique en France et principalement chez les Normans*, 2 vols. (Paris 1920).
Apeldoorn, Lambertus Johannes van, *De kerkelijke goederen in Friesland. Beschrijving van de ontwikkeling van het recht omtrent de kerkelijke goederen in Friesland tot 1795*, 2 vols. (Leeuwarden 1915).
Arnoux, Mathieu and Pierre Monnet (eds.), *Le technicien dans la cité en Europe Occidentale, 1250–1650* (Rome 2004).
Arrault, Alain and Catherine Jami (eds.), *Science and technology in China: a post-Needham perspective* (Turnhout 2001).
Arroyo Ruiz-Zorrilla, Ricardo, *Apunte para una historia de la enseñanza de la nautica en España* (Madrid 1989).
Artz, Frederick B., 'L'éducation technique en France au XVIIIe siècle (1700–1789)', *Revue d'Histoire Moderne*, 13 (1938) 361–347.
Ashtor, Eliyahu, 'The factors of technological and industrial progress in the Later Middle Ages', *The Journal of European Economic History*, 18 (1989) 7–36.
——, *Technology, industry and trade. The Levant versus Europe, 1250–1500* (Hampshire 1992).
Attfield, Robin, 'Christian attitudes to nature', *Journal of the History of Ideas*, 44 (1983) 369–386.
Baar, Marinus de, *Order, change and chance in the European perspective on nature (1600–1800)* (Groningen 2007).
Baetens, Roland, *De nazomer van Antwerpens welvaart. De diaspora van het handelshuis De Groote tijdens de eerste helft der 17[de] eeuw*, 2 vols. (Brussels 1976).
Bailey, William, *The advancement of arts, manufactures and commerce or description of the useful machines and models contained in the repository of the Society of Arts*, 2 vols. (London 1772).

Baldaeus, Philippus, *Naauwkeurige beschryvinge van Malabar en Choromandel, der zelver aangrenzende ryken, en het machtige eyland Ceylon* (Amsterdam 1672).
Barbalet, Jack, *Weber, passion and profits. The Protestant Ethic and the spirit of capitalism* (Cambridge 2008).
Barrera-Osorio, Antonio, *Experiencing nature. The Spanish American empire and the early Scientific Revolution* (Austin, 2006).
Barro, Robert J. and Rachel M. McCleary, 'Religion and economic growth across countries', *American Sociological Review*, 68 (2003) 760–781.
Bartlett, Robert, *The making of Europe. Conquest, colonization and cultural change 950–1350* (London 1993).
——, *The natural and the supernatural in the Middle Ages* (Cambridge 2008).
Baten, Jörg and Jan Luiten van Zanden, 'Book production and the onset of economic growth', *Journal of Economic Growth*, 13 (2008) 217–235.
Bavel, Bas van, *Goederenverwerving en goederenbeheer van de abdij Mariënweerd (1129–1592)* (Hilversum 1993).
Bavel, Bas van, Tine de Moor and Jan Luiten van Zanden, 'The institutional organization of land markets: introduction', *Continuity and Change*, 23 (2008) 9–11.
Beal, Samuel (ed.), *Buddhist records of the Western World*, 2 vols. (London 1906).
Beales, Derek, *Prosperity and plunder. European Catholic monasteries in the age of Revolution, 1650–1815* (Cambridge 2003).
Becker, Sascha O. and Ludger Woessmann, 'Was Weber wrong? A human capital theory of Protestant economic history', *Quarterly Journal of Economics*, 124 (2009) 531–596.
Beeck Calkoen, Jan Frederik van, *Onderzoek naar den rechtstoestand der geestelijke en kerkelijke goederen in Holland na de Reformatie* (Amsterdam 1910).
Belfanti, Carlo Marco, 'Guilds, patents, and the circulation of technical knowledge: Northern Italy during the Early Modern Age', *Technology and Culture*, 45 (2004) 569–589.
Bell, Catherine, ' "A precious raft to save the world". The interaction of scriptural traditions and printing in a Chinese morality book', *Late Imperial China*, 17 (1996) 158–199.
Benedict, Philip, *Rouen during the Wars of Religion* (Cambridge 1981).
——, 'The historiography of Continental Calvinism', in: Hartmut Lehmann and Guenther Roth (eds.), *Weber's Protestant Ethic. Origins, evidence, context* (Cambridge 1983) 305–325.
Benz, Ernest, 'I fondamenti cristiani della technica occidentale', in: Enrico Castelli (ed.), *Tecnica e casistica* (Rome 1964) 241–263.
——, *Schöpfungsgeschichte und Endzeiterwartung* (Munich 1965).
Berg, Christa et al. (eds.), *Handbuch der deutsche Bildungsgeschichte*, 2 vols. (Munich 1996–2005).
Berg, Maxine, *The age of manufactures 1700–1820. Industry, innovation and work in Britain* (2nd. edn., London/New York 1994).
——, 'In pursuit of luxury: Global history and British consumer goods in the eighteenth century', *Past and Present*, number 182 (2004) 85–142.
——, 'Britain, industry and perceptions of China: Matthew Boulton, "useful knowledge" and the Macartney Embassy to China 1792–1794', *Journal of Global History*, 1 (2006) 269–288.
——, 'The genesis of "useful knowledge" ', *History of Science*, 45 (2007) 123–133.
Berg, Johannes van den, 'Orthodoxy, rationalism and the world in eighteenth-century Holland', in: Derek Baker (ed.), *Sanctity and secularity: the Church and the world* (Oxford 1973) 173–192.
Bergin, Joseph, *Church, society and religious change in France* (New Haven 2009).
Bergvelt, Ellinoor and Renée Kistemaker (eds.), *De wereld binnen handbereik. Nederlandse kunst- en rariteitenverzamelingen, 1585–1735* (Zwolle 1992).
Berkel, Klaas van, *Citaten uit het boek der natuur. Opstellen over Nederlandse wetenschapsgeschiedenis* (Amsterdam 1998).

Berman, Harold J., *Law and revolution. The formation of the western legal tradition* (Cambridge Mass. 1983).
Biagioli, Mario, 'From print to patents. Living on instruments in early modern Europe', *History of Science*, 44 (2006) 139–186.
Bicker, Lambertus, 'Rivierkundige grondwaarheden bijzonderlijk toegepast op de rivieren onzes lands tot herstelling derzelven', *Verhandelingen van het Bataafsch Genootschap der Proefondervindelijke Wijsbegeerte*, 1, 1774, 1–210.
Bijker, Wiebe E., Thomas P. Hughes and Trevor Pinch (eds.), *The social construction of technological systems. New directions in the sociology and history of technology* (Cambridge Mass. 1987).
Binding, Guenther, *Baubetrieb im Mittelalter* (Darmstadt 1993).
Bin Wong, Roy, *China transformed. Historical change and the limits of European experience* (Ithaca 1997).
Black, Christopher, *Italian confraternities in the sixteenth century* (Cambridge 1989).
Blair, Ann, 'Afterword: Rethinking Western printing with Chinese comparisons', in: Lucille Chia and Hilde de Weerdt (eds.), *Knowledge and text production in an age of print: China, 900–1400* (Leiden 2011) 349–360.
Bloch, Marc, 'Avènement et conquêtes du moulin à eau', *Annales d'histoire économique et sociale*, 7 (1935) 538–563.
Blum, Ulrich and Leonard Dudley, 'Religion and economic growth: Was Weber right?', *Journal of Evolutionary Economics*, 11 (2001) 207–230.
——, 'Standardised Latin and medieval economic growth', *European Review of Economic History*, 7 (2003) 213–238.
Blussé, Leonard, *Strange company. Chinese settlers, Mestizo women and the Dutch in VOC Batavia* (Dordrecht 1986).
Bodde, Derek, *Chinese thought, society and science. The intellectual and social background of science and technology in pre-modern China* (Honolulu 1991).
Boedeker, Hans Erich, '"Sehen, horen, sammeln und schreiben". Gelehrte Reisen im Kommunikationsystem der Gelehrtenrepublik', *Paedagogica Historica* 38 (2002) 505–532.
Böhm, Laetitia and Charlotte Schönbeck (eds.), *Technik und Bildung* (Düsseldorf 1989).
Böhm, Laetitia, 'Technische Bildung von den Anfängen bis zur frühen Neuzeit', in: Laetitia Böhm and Charlotte Schönbeck (eds.), *Technik und Bildung* (Düsseldorf 1989) 59–117.
Booy, Engelina P. de and Johannes Engel, *Van erfenis tot studiebeurs. De Fundatie van de vrijvrouwe van Renswoude te Delft* (Delft 1985).
Borgato, Maria Teresa, 'Mathematical research in Italian universities in the modern era', in: Mordechai Feingold and Víctor Navarro-Brotóns (eds.), *Universities and science in the early modern period* (Dordrecht 2006) 127–140.
Bots, Jan, *Tussen Descartes en Darwin. Geloof en natuurwetenschap in de achttiende eeuw in Nederland* (Assen 1972).
Bouchard, Constance Brittain, *Holy entrepreneurs. Cistercians, knights, and economic exchange in twelfth-century Burgundy* (Ithaca 1991).
Bouza, Fernando, *Comunicación, conocimiento y memoria en la España de los siglos XVI y XVII* (Salamanca 1999).
——, *Corre manuscrito. Una historia cultural del Siglo de Oro* (Madrid 2001).
Braudel, Fernand, *Civilisation matérielle. Economie et capitalisme XVe–XVIIIe siècle*, 3 dln. (Paris 1979).
Bray, Francesca, *The rice economies. Technology and development in Asian societies* (Oxford 1986).
——, *Technology and gender: Fabrics of power in Late Imperial China* (Berkeley 1997).
——, *Technology and society in Ming China (1368–1644)* (Washington 2000).
——, 'Introduction: The powers of *Tu*', in: Francesca Bray, Vera Dorofeeva-Lichmann and Georges Métailié (eds.), *Graphics and texts in the production of technical knowledge in China. The warp and the weft* (Leiden 2007) 1–78.

——, 'Agricultural illustrations: Blueprint or icon', in: Francesca Bray, Vera Dorofeeva-Lichmann and Georges Métailié (eds.), *Graphics and texts in the production of technical knowledge in China. The warp and the weft* (Leiden 2007) 521–567.

——, 'Science, technique, technology. Passages between matter and knowledge in imperial Chinese agriculture', *British Journal for the History of Science*, 41 (2008) 319–344.

Bray, Francesca, Vera Dorofeeva-Lichmann and Georges Métailié (eds.), *Graphics and texts in the production of technical knowledge in China. The warp and the weft* (Leiden 2007).

Brewer, John and Roy Porter (eds.), *Consumption and the world of goods* (London 1993).

Brianta, Donata, 'Education and training in the mining industry, 1750–1860: European models and the Italian case', *Annals of Science*, 57 (2000) 267–300.

Briels, Johannes G.C.A., *Zuid-Nederlanders in de Republiek 1572–1630. Een demografische en cultuurhistorische studie* (Sint Niklaas 1985).

Brink, Paul van den, *'In een opslag van het oog'. De Hollandse rivierkartografie en waterstaatszorg in opkomst, 1725–1754* (Alphen aan den Rijn 1998).

Broadberry, Stephen and Bishnupriya Gupta, 'The early modern Great Divergence. Wages, prices and economic development in Europe and Asia, 1500–1800', *Economic History Review*, 59 (2006) 2–31.

Broadberry, Stephen and Kevin H. O'Rourke, 'Introduction', in: idem (eds.), *The Cambridge economic history of modern Europe*, vol. 1, *1700–1870* (Cambridge 2010) 1–4.

Brockey, Liam Matthew, *Journey to the East. The Jesuit mission to China 1579–1724* (Cambridge, Mass. 2007).

Brockliss, Laurence W.B., *French higher education in the seventeenth and eighteenth centuries. A cultural history* (Oxford 1987).

Brokaw, Cynthia, *Commerce in culture. The Sibao book trade in the Qing and Republican periods* (Cambridge Mass. 2007).

Brooke, John Hedley, *Science and religion. Some perspectives* (Cambridge 1991).

Brook, Timothy, 'Weber, Mencius, and the history of Chinese capitalism', *Asian Perspective*, 19.1 (1995) 79–97.

——, 'Edifying knowledge. The building of school libraries in Ming China', *Late Imperial China*, 17 (1996) 93–119.

——, *The confusions of pleasure. Commerce and culture in Ming China* (Berkeley 1998).

Bruce, Steve, *Religion in the modern world. From cathedrals to cults* (Oxford 1996).

Bruchhäuser, Hanns-Peter, 'Berufsbildung', in: Berg, Christa et al. (eds.), *Handbuch der deutsche Bildungsgeschichte*, bd.II, ed. by Notker Hammerstein and Ulrich Hammerstein (Munich 2005) 401–419.

Brunings, Christiaan, 'Antwoord op de vraag...: Is de algemeen grondregel der hydrometrie... insgelyks toepasselyk op de zeeboezems, gelyk het Ye...', *Verhandelingen uitgegeeven door de Hollandsche Maatschappij der Wetenschappen*, 24 (1787) 1–58.

Bruton, Eric, *Clock and watches 1400–1900* (New York/Washington 1967).

Bryant, Joseph M., 'The West and the Rest revisited: Debating capitalist origins, European colonialism, and the advent of modernity', *Canadian Journal of Sociology*, 31 (2006) 403–444.

Bunge, Wiep van, 'Introduction', in: Wiep van Bunge (ed.), *The early Enlightenment in the Dutch Republic, 1650–1750* (Leiden 2003) 1–18.

Buonora, Paolo, 'Cartografia e idraulica del Tevere (secoli XVI–XVII)', in: Alessandra Fiocca, Daniela Lamberini and Cesare Maffioli (eds.), *Arte e scienza delle acque nel Rinascimento* (Venice 2003) 169–193.

Buringh, Eltjo and Jan Luiten van Zanden, 'Charting the "Rise of the West". Manuscripts and printed books in Europe. A long-term perspective from the sixth to the eighteenth centuries', *The Journal of Economic History*, 69 (2009) 409–445.

Burke III, Edmund, 'Islam at the center. Technological complexes and the roots of modernity', *Journal of World History*, 20 (2009) 167–186.

Burke, Peter, *A social history of knowledge from Gutenberg to Diderot* (Cambridge 2000).

Bumbacher, Stephan Peter, 'On the pre-Tang Daoist monastic establishments at Mao Shan, according to Daoxue zhuan', *Journal of Chinese Religions*, 28 (2000) 145–160.
Cameron, Alan, *The last pagans of Rome* (Oxford 2011).
Cañizares-Esguerra, Jorge, *How to write the history of the New World. Histories, epistemologies, and identities in the eighteenth-century Atlantic World* (Stanford 2001).
——, *Nature, empire, and nation. Explorations of the history of science in the Iberian world* (Stanford, 2006).
Carls, Wieland (ed.), *Felix Fabri, Die Sionpilger* (Berlin 1999).
Carter, Patrick R.N., 'The fiscal Reformation: clerical taxation and opposition in Henrician England', in: Kümin, Beat A. (ed.), *Reformations old and new. Essays on the socio-economic impact of religious change c.1470–1630* (Aldershot 1996) 92–105.
Cavalcanti, Tiago V., Stephen L. Parente and Rui Zhao, 'Religion in macroeconomics: a quantitative analysis of Weber's thesis', *Economic Theory*, 32 (2007) 105–124.
Cazzola, Franco, 'Le bonifiche cinquecentesche nella valle del Po: governare le acque, creare nuova terra', in: Alessandra Fiocca, Daniela Lamberini and Cesare Maffioli (eds.), *Arte e scienza delle acque nel Rinascimento* (Venice, 2003) 15–35.
Chaffee, John W., *The thorny gates of learning in Sung China: a social history of examinations* (Cambridge 1985).
Chan, Wing-tsit, 'Chu Hsi and the academies', in: Wm.Theodore De Bary and John W. Chaffee (eds.), *Neo-Confucian education: The formative stage* (Berkeley 1989) 389–413.
Chao, Kang, *The development of cotton textile production in China* (Cambridge Mass. 1977).
Chartier, Roger, Dominique Julia, and Marie-Madeleine Compère, *L'éducation en France du XVIe au XVIIIe siècle* (Paris 1976).
——, *Inscrire et effacer. Culture écrite et littérature (XIe–XVIIIe siècle)* (Paris 2000).
Chia, Lucille, '*Mashaben*: Commercial publishing in Jianyang from the Song to the Ming', in: Paul J. Smith and Richard von Glahn (eds.), *The Song-Yuan-Ming transition in Chinese history* (Cambridge 2003) 284–328.
——, *Printing for profit: the commercial publishers of Jianyang, Fujian (11th–17th centuries)* (Cambridge Mass. 2002).
——, 'The uses of print in early Quanzhen Daoist texts', in: Lucille Chia and Hilde de Weerdt (eds.), *Knowledge and text production in an age of print: China, 900–1400* (Leiden 2011) 167–213.
Chia, Lucille and Hilde de Weerdt (eds.), *Knowledge and text production in an age of print: China, 900–1400* (Leiden 2011).
——, 'Introduction', in: Lucille Chia and Hilde de Weerdt (eds.), *Knowledge and text production in an age of print: China, 900–1400* (Leiden 2011).
Cho, Philip S., 'Sericulture songs and the urbanization of silkworm deity temples in 18th century Zhejiang', in; Special issue *The Chinese Journal for the History of Science and Technology* (forthcoming).
Chow, Kai-Wing, *Publishing, culture, and power in early modern China* (Stanford 2004).
Cipolla, Carlo M., *Clocks and culture, 1300–1700* (New York 1967).
Ciriacono, Salvatore, *Acque e agricoltura. Venezia, l'Olanda e la bonifica europea in età moderna* (Milan 1994).
Clossey, Luke, *Salvation and globalization in the early Jesuit missions* (Cambridge 2008).
Clunas, Craig, *Superfluous things. Material culture and social status in early modern China* (Honolulu 1991).
——, *Pictures and visuality in early modern China* (Princeton 1997).
Cohen, Floris, *The Scientific Revolution. A historiographical inquiry* (Chicago 1994).
——, *De herschepping van de wereld. Het ontstaan van de moderne natuurwetenschap verklaard* (Amsterdam 2007).
——, *How modern science came into the world. Four civilizations, one 17th-century breakthrough* (Amsterdam 2012).

Cohen, Patricia Cline, *A calculating people. The spread of numeracy in early America* (2nd edn., London 1999).

——, 'Reckoning with commerce: numeracy in eighteenth-century America', in: John Brewer and Roy Porter (eds.), *Consumption and the world of goods* (London 1993) 320–334.

Cook, Harold J., *Matters of exchange. Commerce, medicine, and science in the Dutch Golden Age* (New Haven 2007).

Cookson, Gillian, 'Millwrights, clockmakers and the origins of textile machine-making in Yorkshire', *Textile History*, 27 (1996) 43–57.

Cooper, Michael (ed.), *João Rodrigues's account of sixteenth-century Japan* (Hakluyt Society, third series, vol. VII, London 2001).

Coşgel, Metin and Lanse Minkler, 'Rationality, integrity, and religious behavior', *Journal of Socio-Economics*, 33 (2004) 329–341.

Cotte, Michel (ed.), *Circulations techniques en amont de l'innovation: hommes, objets et idées en mouvement* (Belfort 2004).

Cressy, David, 'Literacy in context: meaning and measurement in early modern England', in: John Brewer and Roy Porter (eds.) *Consumption and the world of goods* (London 1993) 305–319.

Crosby, Alfred, *The measure of reality: Quantification and Western society, 1250–1600* (Cambridge 1997).

Crossley, Pamela Kyle, 'Manchu education', in: Benjamin A. Elman and Alexander Woodside (eds.), *Education and society in Late Imperial China, 1600–1900* (Berkeley 1994) 340–378.

Crowston, Clare, 'From school to workshop: Pre-training and apprenticeship in Old Regime France', in: Bert De Munck, Steven Kaplan and Hugo Soly (eds.), *Learning on the shop floor. Historical perspectives on apprenticeship* (New York/Oxford 2007) 46–62.

Csikszentmihalyi, Mark, 'Traditional taxonomies and revealed texts in the Han', in: Livia Kohn and Harold Roth (eds.), *Daoist identity: History, lineage, and ritual* (Honolulu 2002) 81–101.

Csikszentmihalyi, Mark and Michael Nylan, 'Constructing lineages and inventing traditions through exemplary figures in early China', *T'oung Pao*, 89 (2003) 59–99.

Cullen, Christopher, 'The science/technology interface in seventeenth-century China', *Bulletin of the School of Oriental and African Studies*, 53 (1990) 295–318.

Cuomo, Serafina, *Technology and culture in Greek and Roman Antiquity* (Cambridge 2007).

Curtis, Emily Byrne, *Glass exchange between Europe and China, 1550–1800. Diplomatic, mercantile and technological interactions* (Aldershot 2009).

DaCosta Kaufmann, Thomas, *Court, cloister and city. The art and culture of Central Europe 1450–1800* (London 1995).

Daelemans, Bert, 'Pieter Huyssens S.J. (1577–1637), an underestimated architect and engineer', in: Piet Lombaerde (ed.), *Innovation and experiment in the early Baroque in the Spanish Netherlands* (Turnhout 2008) 41–52.

Dampnier, Bernard and Paola Vismara (eds.), *Confrèries et dévotions dans la catholicité moderne (mi XVe–début XIXe siècle)* (Rome 2008).

Darnton, Robert, *The business of Enlightenment. A publishing history of the* Encyclopédie *1775–1800* (Cambridge Mass. 1979).

Daston, Lorraine, 'The nature of nature in early modern Europe', *Configurations*, 6.2 (1998) 149–172.

Davids, Karel, 'De migratiebeweging in Leiden in de achttiende eeuw', in: Herman A. Diederiks, Dirk.Jaap Noordam and Heiko D. Tjalsma (eds.), *Armoede en sociale spanning. Sociaal-historische studies over Leiden in de achttiende eeuw* (Hilversum 1985) 137–156.

——, *Zeewezen en wetenschap. De wetenschap en de ontwikkeling van de navigatietechniek in Nederland tussen 1585 en 1815* (Amsterdam/Dieren 1986).

——, 'Universiteiten, illustre scholen en de verspreiding van technische kennis in Nederland, eind 16e–begin 19e eeuw', *Batavia Academica*, 8 (1990) 3–34.
——, 'Ondernemers in kennis. Het zeevaartkundig onderwijs in de Republiek gedurende de zeventiende eeuw', *De zeventiende eeuw*, 7 (1991) 37–48.
——, 'Shifts of technological leadership in early modern Europe', in: Karel Davids and Jan Lucassen (eds.), *A miracle mirrored. The Dutch Republic in European perspective* (Cambridge 1995) 338–366.
——, 'From De la Court to Vreede. Regulation and self-regulation in Dutch economic discourse between 1660 and the Napoleonic Era', *Journal of European Economic History*, 30 (2001) 245–289.
——, 'Amsterdam as a centre of learning in the Golden Age of the Dutch Republic, 1580–1700', in: Patrick Karl O'Brien, Derek Keene, Marjolein C. 't Hart and Herman van der Wee (eds.), *Urban achievement in Early Modern Europe. Golden Ages in Antwerp, Amsterdam and London* (Cambridge 2001) 305–325.
——, 'Innovations in windmill technology in Europe, 1500–1800', *Atti XXXIV Settimana di Studi Economia ed energia, secc. XIII–XVIII. Economia e energia secc. XIII–XVIII* (Prato 2003) 271–291.
——, 'Guilds, guildsmen and technological innovation in early modern Europe: the case of the Dutch Republic' (Working paper 2003, www.lowcountries.nl).
——, 'River control and the evolution of knowledge: a comparison between regions in China and Europe, c.1400–1850', *Journal of Global History*, 1 (2006) 59–79.
——, 'Apprenticeship and guild control in the Netherlands, c.1450–1800', in: Bert De Munck, Steven Kaplan and Hugo Soly (eds.), *Learning on the shop floor. Historical perspectives on apprenticeship* (New York/Oxford 2007) 65–84.
——, *The rise and decline of Dutch technological leadership. Technology, economy and culture in the Netherlands, 1350–1800* (Leiden 2008).
——, 'Dutch and Spanish global networks of knowledge in the early modern period: Structures, connections, changes', in: Lissa Roberts (ed.), *Centres and cycles of accumulation in and around the Netherlands during the Early Modern Period* (Münster 2011) 29–52.
——, 'In the shadow of Jesuits: Vossius and geography', in: Eric Jorink and Dirk van Miert (eds.), *Isaac Vossius (1618–1689) between science and scholarship* (Leiden 2012) 189–206.
Davids, Karel and Jan Lucassen (eds.), *A miracle mirrored. The Dutch Republic in European perspective* (Cambridge 1995).
——, 'Introduction', in: Karel Davids and Jan Lucassen (eds.), *A miracle mirrored. The Dutch Republic in European perspective* (Cambridge 1995) 1–26.
Davies, Margaret Gay, *The enforcement of English apprenticeship. A study in applied mercantilism 1563–1642* (Cambridge Mass. 1956).
Davis, Edward L., *Society and the supernatural in Song China* (Honolulu 2001).
Davis, Natalie Zemon, *Society and culture in early modern France* (Stanford 1975).
Dawson, Christopher (ed.), *The Mongol Mission. Narratives and letters of the Franciscan missionaries in Mongolia and China in the thirteenth and fourteenth centuries* (London/New York 1955).
Dawson, R., *The Chinese chameleon. An analysis of European conceptions of Chinese civilization* (Oxford 1967).
Dean, Kenneth, *Taoist ritual and popular cults of Southeast China* (Princeton 1993).
——, *Lord of the three in one: The spread of a cult in Southeast China* (Princeton 1998).
De Bary, Wm.Theodore, 'Chu Hsi's aims as an educator', in: Wm.Theodore De Bary and John W. Chaffee (eds.), *Neo-Confucian education: The formative stage* (Berkeley 1989) 186–218.
De Bary, Wm.Theodore and John W. Chaffee (eds.), *Neo-Confucian education: The formative stage* (Berkeley 1989).
——, 'Introduction', in: Wm.Theodore De Bary and John W. Chaffee (eds.), *Neo-Confucian education: The formative stage* (Berkeley 1989) 1–15.

Decker, Michael, 'Plants and progress. Rethinking the Islamic Agricultural Revolution', *Journal of World History*, 20 (2009) 187–206.

Degryse, Karel, *De Antwerpse fortuinen: kapitaalsaccumulatie, -investering en –rendement te Antwerpen in de 18de eeuw* (Antwerp 2005).

Delacroix, Jacques and François Nielsen, 'The beloved myth: Protestantism and the rise of industrial capitalism in nineteenth-century Europe', *Social Forces*, 80 (2003) 509–553.

De Munck, Bert, Piet Lourens and Jan Lucassen, 'The establishment and distribution of craft guilds in the Low Countries, 1000–1800', in: Maarten Prak, Catharina Lis, Jan Lucassen and Hugo Soly (eds.), *Craft guilds in the Early Modern Low Countries. Work, power, and representation* (Aldershot 2006) 32–73.

De Munck, Bert, Steven Kaplan and Hugo Soly (eds.), *Learning on the shop floor. Historical perspectives on apprenticeship* (New York/Oxford 2007).

——, '"Learning on the shop floor" in historical perspective', in: idem, *Learning on the shop floor. Historical perspectives on apprenticeship* (New York/London 2007) 3–32.

De Munck, Bert, *Technologies of learning. Apprenticeship in Antwerp guilds from the 15th century to the end of the Ancien Régime* (Turnhout 2007).

Deng, Kent, *Development versus stagnation. Technological continuity and agricultural progress in premodern China* (Westport, Conn. 1993).

——, 'A critical survey of recent research in Chinese economic history', *Economic History Review*, 53 (2000) 1–28.

Dillon, Michael, 'Jingdezhen as a Ming industrial center', *Ming Studies*, 6 (1978) 37–44.

——, 'Transport and marketing in the development of the Jingdezhen porcelain industry during the Ming and Quing dynasties', *Journal of the Economic and Social History of the Orient*, 35 (1992) 278–290.

Dinis, Alfredo, 'Giovanni Battista Riccioli and the science of his times', in: Mordechai Feingold (ed.), *Jesuit science and the Republic of Letters* (Cambridge Mass. 2003) 195–224.

Dodgen, Randall A., *Controlling the Dragon. Confucian engineers and the Yellow River in Late Imperial China* (Honolulu 2001).

Dohrn-Van Rossum, Gerhard, 'Migration-Innovation-Städtenetze: Ingenieure und technische Experten', in: Matthieu Arnoux and Pierre Monnet (eds.), *Le technicien dans la cité dans en Europe Occidentale, 1250–1650* (Rome 2004) 291–307.

Donnelly, John Patrick and Michael W. Maker (eds.), *Confraternities and Catholic reform in Italy, France and Spain* (Kirksalle 1999).

Drayton, Richard, *Nature's government. Science, imperial Britain, and the "improvement" of the world* (New Haven 2000).

Duchesne, Ricardo, 'On the rise of the West: Researching Kenneth Pomeranz's Great Divergence', *Review of Radical Political Economics*, 36 (2004) 52–81.

Dudley, Leonard, *Information revolutions in the history of the West* (Cheltenham 2008).

Dülmen, Richard von and Sina Rauschenbach (eds), *Macht des Wissens. Die Entstehung der modernen Wissensgesellschaft* (Cologne 2004).

Duffy, Eamon, *The stripping of the altars. Traditional religion in England 1400–1580* (2nd edn., New Haven 2005).

Dux, Günter, 'The genesis of philosophy in the history of mind: A cross-cultural comparison between classical Greece and China', in: Hans Ulrich Vogel and Günter Dux (eds.), *Concepts of nature. A Chinese-European cross-cultural perspective* (Leiden 2010) 103–135.

Dyke, Paul A. van, *The Canton trade. Life and enterprise on the China Coast 1700–1845* (Hong Kong 2005).

Eberhard, Wolfram, *Guilt and sin in traditional China* (Berkeley/Los Angeles 1967).

Ebrey, Patricia and Peter Gregory (eds.), *Religion and society in T'ang and Sung China* (Honolulu 1993).

Ebrey, Patricia, *The five "Confucian" classics* (New Haven 2001).

Echávarria Otero, Javier et al., 'The Twenty-One Books of Devices and Machines: An encyclopedia of machines and mechanisms of the 16th century', in: Hong-Sen Yan and Marco Ceccarelli (eds.), *International symposium on history of machines and mechanisms Proceedings of HMM 2008* (s. l. 2009) 115–132.

Eck, Ton van, 'Geschiedenis van de orgelbouw in de overwegend rooms-katholieke landen en streken', in: Anton Vernooij (ed.), *Van Satans Fluytencast tot Vox Angelica. Handboek voor de kerkorganist* (Kampen 2003) 117–272.

Egan, Ronald, 'To count grains of sand on the ocean floor: Changing perceptions of books and learning in the Song dynasty', in: Lucille Chia and Hilde de Weerdt (eds.), *Knowledge and text production in an age of print: China, 900–1400* (Leiden 2011) 33–62.

Eggert, Marion, *Vom Sinn des Reisens. Chinesische Reiseschriften vom 16. bis zum frühen 19. Jahrhundert* (Wiesbaden 2006).

Ehmer, Josef, 'Artisans, journeymen, guilds, and labor markets: Thinking about European comparative perspectives', in: Paolo Massa and Angelo Moioli (eds.), *Dalla corporazione al mutuo soccorso. Organizazione e tutela del lavoro tra XVI e XX secolo* (Milan 2006) 57–69.

Eisenstein, Elisabeth, *The printing press as an agent of change. Communications and cultural transformations in early-modern Europe* (Cambridge 1979).

Ellul, Jacques, 'The relation of Man to Creation according to the Bible', in: Carl Mitcham and Jim Grote (eds.), *Theology and technology. Essays in Christian analysis and exegesis* (Lanham 1984) 139–155.

——, 'Technique and the opening chapters of Genesis', in: Carl Mitcham and Jim Grote (eds.), *Theology and technology. Essays in Christian analysis and exegesis* (Lanham 1984) 123–137.

Elman, Benjamin A., 'Changes in Confucian civil service examinations from the Ming to the Ch'ing dynasty', in: Benjamin A. Elman and Alexander Woodside (eds.), *Education and society in Late Imperial China, 1600–1900* (Berkeley 1994) 111–149.

——, *From philosophy to philology. Social and intellectual aspects of change in Late Imperial China* (2nd edn., Los Angeles 2001).

——, *A cultural history of civil examinations in Late Imperial China* (Berkeley 2000).

——, *On their own terms. Science in China, 1550–1900* (Cambridge Mass. 2005).

Elman, Benjamin A. and Alexander Woodside (eds.), *Education and society in Late Imperial China, 1600–1900* (Berkeley 1994).

Elvin, Mark, *The pattern of the Chinese past* (London 1973).

——, 'Why China failed to create an endogenous industrial capitalism: A critique of Max Weber's explanation', *Theory and Society*, 13 (1984) 379–391.

——, 'The high-level equilibrium trap: The causes of the decline of invention in the traditional Chinese textile industries', in: William E. Willmott, *Economic organization in Chinese Society* (Stanford 1972) 137–172.

——, *The retreat of the elephants. An environmental history of China* (New Haven 2004).

——, 'Overview', in: Hans Ulrich Vogel and Günter Dux (eds.), *Concepts of nature. A Chinese-European cross-cultural perspective* (Leiden 2010) 1–55.

Elvin, Mark and Ts'ui-jung Liu (eds.), *Sediments of time. Environment and society in Chinese history* (Cambridge 1998).

Emigh, Rebecca J., 'Numeracy or enumeration? The uses of numbers by states and society', *Social Science History*, 26 (2002) 653–698.

Emiliani, Cesare and Shale Niskin, 'Letter to the Editor', *Science* 156 (12 May 1967), 738.

Endres, Rudolf, 'Handwerk – Berufsbildung', in: Christa Berg et al. (eds.), *Handbuch der deutsche Bildungsgeschichte*, bd.I, ed. by Notker Hammerstein (Munich 1996) 375–424.

Engelfriet, Peter M., *Euclid in China. The genesis of the first Chinese translation of Euclid's Elements books I–VI (Jihe yanben; Beijing 1627) and its reception up to 1723* (Leiden 1998).

Epstein, Larry, 'Journeymen mobility and the circulation of technical knowledge, XVI–XVIIth centuries', in: Liliane Hilaire-Pérez and Anne-Françoise Garçon (eds.), *Les chemins de la nouveauté. Innover, inventer au regard de l'histoire* (Paris 2003) 411–430.

——, 'Labour mobility, journeyman organizations and markets in skilled labour in Europe, 14th–18th centuries', in: Mathieu Arnoux and Pierre Monnet (eds.), *Le technicien dans la cité en Europe Occidentale, 1250–1650* (Rome 2004) 251–269.

Epstein, Larry and Maarten Prak (eds.), *Guilds, innovation and the European economy, 1400–1800* (Cambridge 2008).

Epstein, Steven, *Wage labor and guilds in medieval Europe* (Chapel Hill/London 1995).
Eschenlohr, Ludwig, 'Les circulations techniques au Moyen Âge: l' éclairage de l'archéologie dans le domaine de la sidérurgie jurassienne', in: Michel Cotte (ed.), *Circulations techniques en amont de l'innovation: hommes, objets et idées en mouvement* (Belfort 2004) 55–71.
Fabri, Ria and Piet Lombaerde, 'Architectural treatises, books and prints in the libraries of the Jesuits in Antwerp', in: Piet Lombaerde (ed.), *Innovation and experiment in the early Baroque in the Spanish Netherlands* (Turnhout 2008) 187–200.
Farr, James A., *Hands of honor. Artisans and their world in Dijon 1550–1650* (Ithaca 1988).
Fassl, Peter, *Konfession, Wirtschaft und Politik. Von der Reichsstadt zur Industriestadt. Augsburg 1750–1850* (Sigmaringen 1988).
Feenstra, Ernest S., 'Christian impact on ecology', *Science*, 156 (3776), 12 May 1967, 737.
Feilke, Herbert, *Felix Fabris Evagatorium über seine Reise in das Heilige Land. Eine Untersuchung über die Pilgerliteratur des ausgehenden Mittelalters* (Bern 1976).
Findlen, Paula, *Possessing nature. Museums, collecting, and scientific culture in early modern Italy* (Berkeley 1994).
——, 'Scientific spectacle in Baroque Rome: Athanasius Kircher and the Roman College Museum', in: Mordechai Feingold (ed.) *Jesuit science and the Republic of Letters* (Cambridge Mass. 2003) 225–284.
Finlay, Robert, 'China, the West, and world history in Joseph Needham's *Science and Civilisation in China*', *Journal of World History*, 11 (2000) 265–303.
Finnane, Antonia, *Speaking of Yangzhou: A Chinese city, 1550–1850* (Cambridge Mass. 2004).
Fiocca, Alessandra, 'Regolamentazione delle acque e transfert tecnologico tardo rinascimento. Il caso di Ferrara e Venezia', in: Alessandra Fiocca, Daniela Lamberini and Cesare Maffioli (eds.), *Arte e scienza delle acque nel Rinascimento* (Venice 2003) 137–167.
Fissell, May and Roger Cooter, 'Exploring natural knowledge. Science and the popular', in: Roy Porter (ed.), *The Cambridge History of Science*, vol. 4 *Eighteenth-century science* (Cambridge 2003) 129–158.
Flessel, K., *Der Huang-ho und die historische Hydrotechnik in China* (Tübingen 1974).
Flint, Valerie I.J., *The rise of magic in early medieval Europe* (Princeton 1991).
Földvári, Peter, Bas van Leeuwen and Jieli van Leeuwen-Li, 'How did women count? A note on gender-specific age heaping difference in the sixteenth to nineteenth centuries', *Economic History Review*, 65 (2012) 303–313.
Forbes, Robert J. 'Power', in: Charles Singer et al. (ed.), *History of Technology*, vol. II (Oxford 1956) 589–622.
Fossier, Robert, 'The rural economy and demographic growth', in: David Luscombe and Jonathan Riley-Smith (eds.), *The New Cambridge Medieval History*, vol. IV part I (Cambridge 2004) 11–46.
Foulk, T. Griffith, 'Myth, ritual, and monastic practice in Sung Ch'an Buddhism', in: Patricia Ebrey and Peter Gregory, *Religion and society in T'ang and Sung China* (Honolulu 1993) 147–208.
France, John, 'Technology and the success of the First Crusade', in: Lev Jaacov (ed.), *War and society in the Eastern Mediterranean, 7th–15th centuries* (Leiden 1996) 163–176.
François, Étienne, *Protestants et catholiques. Identités et pluralisme Augsbourg, 1648–1806* (Paris 1993).
Frank, Andre Gunder, *ReOrient. Global economy in the Asian age* (Berkeley 1998).
Franke, O., *Kêng tschi t'u. Ackerbau und Seidengewinnung in China. Ein kaiserliches Lehr- und Mahnbuch* (Hamburg 1913).
Freedman, Paul, 'The medieval other: The Middle Ages as Other', in: Timothy S. Jones and David A. Sprunger (eds.), *Marvels, monsters and miracles. Studies in the medieval and early modern imaginations* (Kalamazoo 2002).
Friedel, Robert, *A culture of improvement: Technology and the Western millennium* (Cambridge Mass. 2007).

Friedsam, Manfred, 'L'enseignement des mathématiques sous les Song et les Yuan', in: Christine Nguyen Tri and Catherine Despeux (eds.), *Éducation et instruction en Chine: II: Les formations spécialisées* (Paris/Louvain 2003) 49–68.
Friese, Heinz, *Das Dienstleistungs-System der Ming-Zeit (1368–1644)* (Hamburg 1959).
Frisi, Paolo, 'Berigt aan de Hollandsche Maatschappij der Weetenschappen van... Paulo Frisi nopens de verdeeling en zamenloop der rivieren', *Verhandelingen uitgegeeven door de Hollandsche Maatschappij der Wetenschappen*, 14 (1773) 112–130.
Frijhoff, Willem, 'Van onderwijs naar opvoedend onderwijs. Ontwikkelingslijnen van opvoeding en onderwijs in Noord-Nederland in de achttiende eeuw', in: Hans Bots (ed.), *Onderwijs en opvoeding in de achttiende eeuw. Verslag van het symposium, Doesburg 1982* (Amsterdam/Maarssen 1983) 3–40.
Galabert, Juan, 'Castile 1504–1808 ', in: Richard Bonney (ed.), *The rise of the fiscal state in Europe c.1200–1815* (Oxford 1999) 201–242.
Galor, Oded and David N. Weil, 'Population, technology and growth. From the malthusian regime to the demographic transition', *American Economic Review*, 90 (2000) 806–828.
Galuzzi, Paolo, 'Portraits of machines in fifteenth-century Siena', in: L.R. Mazzoloni (ed.), *Non-verbal communication in science prior to 1900* (Florence 1993) 53–90.
——, *Renaissance engineers from Brunelleschi to Leonardo da Vinci* (Florence 1996).
García Garralón, Marta, 'The education of pilots for the Indies trade in Spain during the eighteenth century', *International Journal of Maritime History*, 21 (2009) 189–220.
García Tapia, Nicolás, *Tecnica y poder en Castilla durante los siglos XVI y XVII* (Salamanca 1989).
——, *Ingeniera y arquitectura en el Renacimiento español* (Valladolid 1990).
——, *Un inventor navarro. Jerónimo de Ayanz y Beaumont 1553–1613* (Pamplona 2001).
García Tapia, Nicolás and Jesús Carillo Castillo, *Tecnología y imperio. Ingenios y leyendas del siglo de oro* (Madrid 2002).
Gardella, Robert, 'Squaring accounts. Commercial bookkeeping methods and capitalist rationalism in Late Qing and Republican China', *Journal of Asian Studies*, 51 (1992) 317–339.
Gasparetto, Astone, *Il vetro di Murano dalle origine ad oggi* (Venice 1958).
Gatto, Romano, 'Christoph Clavius' "Ordo Servandus in Addiscendis Disciplinis Mathematicis" and the teaching of mathematics in Jesuit colleges at the beginning of the modern era', *Science and Education*, 15 (2006) 235–258.
Gaukroger, Steven, *The emergence of a scientific culture. Science and the shaping of modernity 1210–1685* (Oxford 2006).
Gelderblom, Oscar, *Zuid-Nederlandse kooplieden en de opkomst van de Amsterdam stapelmarkt (1578–1630)* (Hilversum 2002).
Gellner, David N., 'Max Weber, capitalism, and the religion of India', in: Richard H. Roberts (ed.), *Religion and the transformation of capitalism. Comparative approaches* (London 2002) 21–46.
Gernet, Jacques, *Buddhism in Chinese society. An economic history from the fifth to the tenth centuries* (New York 1995).
Gerritsen, Anne, 'Fragments of a global pasts: Ceramics manufacture in Song-Yuan-Ming Jingdezhen', *Journal of the Economic and Social History of the Orient*, 52 (2009) 117–152.
——, 'Ceramics for local and global markets: Jingdezhen's agora of technologies', in: Dagmar Schäfer (ed.), *Cultures of knowledge. Technology in Chinese history* (Leiden 2012) 161–184.
Gies, Frances and Joseph, *Cathedral, forge and waterwheel. Technology and invention in the Middle Ages* (New York 1994).
Gillispie, Charles Coulston, *Science and polity in France at the end of the Old Regime* (Princeton 1980).
Glacken, Clarence, *Traces on the Rhodian shore. Nature and culture in Western thought from Ancient time to the end of the eighteenth century* (Berkeley 1967).

Glahn, Richard von, 'Towns and temples: Urban growth and decline in the Yangzi Delta, 1100–1400', in: Paul J. Smith and Richard von Glahn (eds.), *The Song-Yuan-Ming transition in Chinese history* (Cambridge 2003) 176–211.
Glick, Thomas F., *Irrigation and society in Medieval Valencia* (Cambridge Mass. 1970).
——, 'Moriscos y marranos come agentes de la diffusion tecnológica', *Arbor* CXLIX 586–587 (1994) 113–131.
——, *Islamic and Christian Spain in the early Middle Ages* (Leiden 2005).
Glick, Thomas F., Steven J. Livesey and Faith Wallis (eds.), *Medieval science, technology, and medicine. An encyclopedia* (New York 2005).
Goeree, Willem, *d'Algemeene bouwkunst* (Amsterdam 1681).
Goff, Jacques le, *Pour un autre Moyen Âge. Temps, travail et culture en Occident* (Paris 1977).
Golas, Peter J., 'Early Ch'ing guilds', in: G. William Skinner (ed.), *The city in Late Imperial China* (Stanford 1977) 555–580.
——, 'Technological illustration in China: A post-Needham perspective', in: Alain Arrault and Catherine Jami (eds.), *Science and technology in China: a post-Needham perspective* (Turnhout 2001) 43–58.
——, '"Like obtaining a great treasure". The illustrations in Song Yingxing's *The exploitation of the works of nature*', in: Francesca Bray, Vera Dorofeeva-Lichmann and Georges Métailié (eds.), *Graphics and texts in the production of technical knowledge in China. The warp and the weft* (Leiden 2007) 569–614.
Goldgar, Anne, *Impolite learning. Conduct and community in the Republic of Letters 1680–1730* (New Haven 1995).
Goldstone, Jack A., 'The rise of the West – or not? A revision to socio-economic history', *Sociological Theory*, 18 (2000) 175–194.
——, 'Efflorescences and economic growth in world history: Rethinking the "rise of the West" and the Industrial Revolution', *Journal of World History*, 13 (2002) 323–389.
Goldthwaite, Richard, 'Schools and teachers of commercial arithmetic in Renaissance Florence', *Journal of European Economic History*, 1 (1972) 418–433.
Golinski, Jan, *Science as public culture. Chemistry and Enlightenment in Britain 1760–1820* (Cambridge 1992).
Goodman, David C., *Power and penury. Government, technology and science in Philip II's Spain* (Cambridge 1988).
Goody, Jack, *Capitalism and modernity. The Great Debate* (Cambridge 2004).
Gorski, Philip 'The Little Divergence. The Protestant Reformation and economic hegemony in early modern Europe', in: William H. Swatos jr. and Lutz Kaelber (eds.), *The Protestant Ethic turns 100. Essays on the Centenary of the Weber Thesis* (Boulder/London 2005) 165–190.
Gottlieb, Roger S. (ed.), *The Oxford handbook of religion and ecology* (Oxford 2006).
Goudswaard, Nicolaas B., *Vijfenzestig jaren nijverheidsonderwijs* (Assen 1981).
Graff, Harvey, *The legacies of literacy. Continuities and contradictions in Western culture and society* (Bloomington 1987).
Grantham, George, 'Contra-Ricardo: on the macro-economics of pre-industrial economics', *European Review of Economic History*, 3 (1999) 199–232.
Greene, Kevin, 'Technology and innovation in context: the Roman background to medieval and later developments', *Journal of Roman Archaeology*, 7 (1994), 22–33.
——, 'Technological innovation and economic progress in the ancient world: M.I. Finley re-considered', *Economic History Review*, 53 (2000) 29–59.
Gregory, Brad S., *The unintended Reformation. How a religious revolution secularized the world* (Cambridge Mass. 2012).
Guo, Qingua, *Chinese architecture and planning. Ideas, methods, techniques* (Stuttgart/London 2005).
Guy, R. Kent, *The emperor's Four Treasuries. Scholars and the state in the Late Ch'ieng-Lung era* (Cambridge 1987).

Hall, Bert S., '"Der Meister sol auch kennen schreiben und lesen". Writings about technology, ca.1400–ca.1600 A.D. and their cultural implications', in: D. Schmandt-Besserat (ed.), *Early technologies* (Malibu 1979) 47–58.
——, 'Lynn White's *Medieval Technology and Social Change* after thirty years', in: Robert Fox (ed.), *Technological change. Methods and themes* (Amsterdam 1996) 85–101.
Halperin, Mark, *Out of the cloister. Literati perspectives on Buddhism in Sung China, 960–1279* (Cambridge Mass. 2006).
Hammerstein, Notker and Rainer A. Müller, 'Das katholische Gymnasialwesen im 17. und 18. Jahhundert', in: Christa Berg et al. (eds.), *Handbuch der deutsche Bildungsgeschichte*, bd.II, ed. by Notker Hammerstein and Ulrich Hammerstein (Munich 2005) 324–354.
Hansen, Valerie, *Changing gods in medieval China, 1127–1276* (Princeton 1990).
Harbsmeier, Christoph, 'Towards a conceptual history of some concepts of nature in classical Chinese: Zì Rán and Zì Rán Zhí Lǐ', in: Hans Ulrich Vogel and Günter Dux (eds.), *Concepts of nature. A Chinese-European cross-cultural perspective* (Leiden 2010) 220–254.
Harris, Steven J., 'Long-distance corporations, big sciences, and the geography of knowledge', *Configurations*, 6 (1998) 269–304.
——, 'Mapping Jesuit science: The role of travel in the geography of knowledge', in: John W. O'Malley et al. (eds.), *The Jesuits. Cultures, sciences, and the arts 1540–1773* (Toronto 1999) 212–240.
——, 'Jesuit scientific activity in overseas missions, 1540–1773', *Isis*, 96 (2005) 71–79.
Hart, Roger, 'The Great Explanandum', *American Historical Review*, 105 (2000) 486–493.
Hartwell, Robert M., 'A cycle of economic change in Imperial China: Coal and iron in northeast China, 750–1350', *Journal of the Economic and Social History of the Orient*, 10 (1967) 102–159.
——, 'Demographic, political, and social transformations of China, 750–1550', *Harvard Journal of Asiatic Studies*, 42 (1982) 365–442.
Haug, Hans et al., *La cathédrale de Strasbourg* (Strasbourg 1957).
Hazewinkel, Hendrik Cornelis, *Geschiedenis van Rotterdam* (2nd edn., Zaltbommel 1974).
Heilbron, John L., *The Sun in the church. Cathedrals as solar observatories* (Cambridge Mass. 1999).
Heirman, Ann and Stephan Peter Bumbacher (eds.), *The spread of Buddhism* (Leiden 2007).
Heller, Henry, *Labour, science and technology in France 1500–1620* (Cambridge 1996).
Hellyer, Marcus, *Catholic physics. Jesuit natural philosophy in early modern Germany* (Notre Dame 2005).
Henderson, John B., 'Cosmology and concepts of nature in traditional China', in: Hans Ulrich Vogel and Günter Dux (eds.), *Concepts of nature. A Chinese-European cross-cultural perspective* (Leiden 2010) 181–197.
Henninger-Voss, Mary, 'Measures of success. Military engineering and the architectonic understanding of design', in: Wolfgang Lefèvre (ed.) *Picturing machines 1400–1700* (Cambridge Mass. 2004) 143–169.
Heuvel, Charles van den, 'Willem Goeree (1635–1711) en de ontwikkeling van een algemene architectuurtheorie in de Nederlanden', *Bulletin KNOB*, 96 (1997) 154–176.
Hilaire-Pérez, Liliane, *L' invention technique au siècle des Lumières* (Paris 2000).
——, 'Les dépôt d'inventions en France et en Angleterre au XVIIIe siècle', in: Patrice Bret, Christiane Demeulenaere and Liliane Hilaire-Pérez (eds.), *Des matériaux pour l' historie. Archives et collections scientifiques et techniques du XVIIe siècle à nos jours* (Paris 2000) 25–38.
Hilaire-Pérez, Liliane and Anne-Françoise Garçon (eds.), *Les chemins de la nouveauté. Innover, inventer au regard de l'histoire* (Paris 2003).
Hilaire-Pérez, Liliane and Cathérine Verna, 'Les circulations techniques: hommes, produits, savoirs au Moyen Âge et à l' époque moderne (Orient, Occident)', in: Michel Cotte (ed.), *Circulations techniques en amont de l'innovation: hommes, objets et idées en mouvement* (Belfort 2004) 11–35.

Hill, Donald R., *Studies in medieval Islamic technology* (Hampshire 1998).
Hinrichs, Ernst, 'Alphabetisierung', in: Dülmen, Richard von and Sina Rauschenbach (eds.), *Macht des Wissens. Die Entstehung der modernen Wissensgesellschaft* (Cologne 2004) 539–561.
Hobson, John M., *The Eastern origins of Western civilisation* (Cambridge 2004).
Hoffman, Philip T., *Growth in a traditional society The French countryside 1450–1815* (Princeton 1996).
Hofmann, Martin, 'The biographer's view of craftmanship', in: Dagmar Schäfer (ed.), *Cultures of knowledge. Technology in Chinese history* (Leiden 2012) 283–297.
Holt, Richard, 'Medieval technology and the historians: the evidence for the mill', in: Robert Fox (ed.), *Technological change: Methods and themes* (Amsterdam 1996) 103–121.
——, 'Mechanization and the medieval English economy' in: Elisabeth Bradford Smith and Michael Wolfe (eds.), *Technology and resource use: cathedrals, mills, mines* (Aldershot 1997) 139–157.
Hoppenbrouwers, Peter and Jan Luiten van Zanden (eds.), *Peasants into farmers? The transformation of rural economy and society in the Low Countries (middle ages–16th century) in light of the Brenner debate* (Turnhout 2001).
——, 'Restyling the transition from feudalism to capitalism. Some critical reflections on the Brenner thesis', in: Peter Hoppenbrouwers and Jan Luiten van Zanden (eds.), *Peasants into farmers? The transformation of rural economy and society in the Low Countries (middle ages–16th century) in light of the Brenner debate* (Turnhout 2001) 19–40.
Hoskins, William G., *The age of plunder. King Henry's England 1500–1547* (London 1976).
Houston, Rab, 'Literacy and society in the West, 1500–1850 ', *Social History*, 8 (1983) 269–293.
Hoven, Birgit van den, *Work in Ancient and Medieval thought. Ancient philosophers, medieval monks and theologians and their concept of work, occupations and technology* (Leiden 1996).
Huang, Philip C.C., *The peasant economy and social change in North China* (Stanford 1985).
——, *The peasant family and rural development in the Yangzi Delta, 1350–1988* (Stanford 1990).
Huang, Shih-shan Susan, 'Early Buddhist illustrated prints in Hangzhou', in: Lucille Chia and Hilde de Weerdt (eds.), *Knowledge and text production in an age of print: China, 900–1400* (Leiden 2011) 135–165.
Hubrig, Hans, *Die patriotische Gesellschaften des 18. Jahhunderts* (Weinheim 1957).
Huey, Herbert, 'European concepts of the Chinese economy: French Jesuits' views on China', *Papers on Far Eastern History*, 31 (1985) 95–116.
Huff, Toby E., *The rise of early modern science: Islam, China, and the West* (Cambridge 1993).
——, *Intellectual curiosity and the Scientific Revolution. A global perspective* (Cambridge 2011).
Humphries, Jane, 'English apprenticeship. A neglected factor in the First Industrial Revolution', in Paul A. David and Mark Thomas (eds.), *The economic future in historical perspective* (Oxford 2003) 73–102.
Hunt, Edward D., *Holy Land pilgrimage in the Later Roman Empire AD 312–460* (Oxford 1982).
Hunt, Noreen (ed.), *Cluniac monasticism in the Central Middle Ages* (Hamden 1971).
Hunter, Michael, *Science and society in Restoration England* (Cambridge 1981).
——, *Science and the shape of orthodoxy. Intellectual change in late-seventeenth century Britain* (Woodbridge 1995).
Huppert, George, *Public schools in Renaissance France* (Urbana/Chicago 1984).
Icher, François, *Les oeuvriers des cathédrales* (sine loco 1998).
Inkster, Ian, *Scientific culture and urbanisation in industrialising Britain* (Ashgate 1998).
——, 'Technology and industrial change: A comparative essay', in: Roy Porter (ed.), *The Cambridge History of Science*, vol. 4 *Eighteenth-century science* (Cambridge 2003) 845–881.

——, 'Potentially global: "Useful and reliable knowledge" and material progress in Europe, 1474–1914 ', *The International History Review*, 28 (2006) 237–286.

Israel, Jonathan I., *Radical Enlightenment. Philosophy and the making of modernity 1650–1750* (Oxford 2001).

Jacob, Margaret, *Scientifc culture and the making of the Industrial West* (New York 1997).

Jacob, Margaret and Matthew Kadane, 'Missing, now found in the eighteenth-century: Weber's Protestant capitalism', *American Historical Review*, 108 (2003) 20–49.

Jäger, Fritz, 'Das Buch von den wunderbaren Maschinen. Ein Kapitel aus der Geschichte der Abendländisch-Chinesischen Kulturbeziehungen', *Asia Major*, 1.1. (New Series) (1944) 78–96.

James, John, *The template makers of the Paris basin* (Leura 1989).

Jami, Catherine, 'Learning mathematical sciences during the Early and Mid Ching', in: Benjamin A. Elman and Alexander Woodside (ed.), *Education and society in Late Imperial China, 1600–1900* (Berkeley 1994) 223–256.

Janssen, Jan A.M.M., *Op weg naar Breda. De opleiding van officieren voor het Nederlandse leger tot aan de oprichting van de Koninklijke Militaire Academie in 1828* (The Hague 1989).

Jensen, Lionel, *Manufacturing Confucianism. Chinese traditions and universal civilization* (Durham 1998).

Jeremy, David J., 'Damming the flood. British government efforts to check the outflow of technicians and machinery 1780–1843', *Business History Review*, 51 (1977) 1–34.

Jixing, Pan, Hans Ulrich Vogel and Elisabeth Theisen-Vogel, 'Die Übersetzung und Verbreitung von Georgius Agricolas "De Re Metallica" im China der späten Ming-Zeit (1368–1644)', *Journal of the Economic and Social History of the Orient*, 32 (1989) 153–202.

Jochum, Uwe, 'Am Ende der Sammlung. Bibliotheken im frühmodernen Staat', in: Richard von Dülmen and Sina Rauschenbach (eds.), *Macht des Wissens. Die Entstehung der modernen Wissensgesellschaft* (Cologne/Weimar/Vienna 2004) 273–294.

Johns, Adrian, *The nature of the book. Print and knowledge in the making* (Chicago/London 1998).

Jörg, Christiaan, *Porcelain and the Dutch China trade* (The Hague 1982).

Jones, William B. and A. Warren Matthews, 'Toward a taxonomy of technology and religion', *Research in Philosophy and Technology*, 10 (1990) 3–23.

Jonkers, Art, 'Rewards and prizes', *The Oxford Encyclopedia of Maritime History*, vol. III (Oxford 2007) 433–436.

Jorink, Eric, *Het Boeck der Natuere. Nederlandse geleerden en de wonderen van Gods Schepping 1575–1715* (Leiden 2006).

Jowett, A.J. and Y. Li, 'Age heaping: contrasting patterns from China', *GeoJournal*, 28 (1992) 427–442.

Kagan, Richard L., *Students and society in early modern Spain* (Baltimore 1974).

Kaplan, Benjamin, *Divided by faith. Religious faith and the practice of toleration in early modern Europe* (Harvard UP 2007).

Kaye, Joel, *Economy and nature in the fourteenth century. Money, market exchange, and the emergence of scientific thought* (Cambridge 1998).

Keene, Derek, 'The early history of English guilds: their role in social and economic organization', in: Paolo Massa and Angelo Moioli (eds.), *Dalla corporazione al mutuo soccorso. Organizazione e tutela del lavoro tra XVI e XX secolo* (Milan 2006) 71–85.

Keller, Evelyn Fox, *Reflections on science and gender* (New Haven 1984).

Kieschnick, John, *The impact of Buddhism on Chinese material culture* (Princeton 2003).

Kirkland, Russell, *Taoism. The enduring tradition* (London 2004).

Klein, Peter W., 'Nederlandse glasmakerijen in de 17[de] en 18[de] eeuw', *Economisch- en Sociaal-Historisch Jaarboek*, 44 (1981) 31–43.

Klever, Wim, *Mannen rond Spinoza (1650–1700). Presentatie van een emanciperende generatie* (Hilversum 1997).

Knowles, David, *The religious orders in England*, vol. II, *The End of the Middle Ages* (Cambridge 1955).

Kohn, Livia, *Monastic life in medieval Daoism* (Honolulu 2003).
Krabbe, Coert Peter, *Ambacht, kunst, wetenschap. De bevordering van de bouwkunst in Nederland (1775–1880)* (Amsterdam 1997).
Kümin, Beat A., 'Reformations old and new: an introduction', in: Beat A. Kümin (ed.), *Reformations old and new. Essays on the socio-economic impact of religious change c.1470–1630* (Aldershot 1996) 1–17.
—— (ed.), *Reformations old and new. Essays on the socio-economic impact of religious change c.1470–1630* (Aldershot 1996).
Kuhn, Dieter, *Die Webstühle des Tzu-jen-i-chih aus der Yüan-Zeit* (Wiesbaden 1977).
——, *The Age of Confucian rule. The Song transformation of China* (Cambridge Mass. 2008).
Kuijpers, Erika, *Migrantenstad. Immigratie en sociale verhoudingen in 17e-eeuws Amsterdam* (Amsterdam 2005).
Lach, Donald and Dale Van Kley, *Asia in the making of Europe*, vol. III, book one (Chicago/London 1993).
Ladurie, E. Le Roy, *Les paysans de Languedoc* (Paris 1966).
Lal, Deepak, *Unintended consequences. The impact of factor endowments, culture, and politics in long-run economic performance* (Cambridge Mass. 1998).
Lamouroux, Christian, 'From the Yellow River to the Huai. New representations of a river network and the hydraulic crisis of 1128', in: Mark Elvin and Ts'ui-jung Liu (eds.), *Sediments of time. Environment and society in Chinese history* (Cambridge 1998) 545–584.
Lane, Frederic C., *Venice. A maritime republic* (Baltimore, 1973).
Landes, David S., *Revolution in time. Clocks and the making of the modern world* (Cambridge Mass. 1983).
——, *The wealth and poverty of nations. Why some are so rich and others are so poor* (London 1998).
Landmann, Salcia, 'Religiöse Bezüge zur Technik im Judentum', in: Ansgar Stöcklein and Mohammed Rassem (eds.), *Technik und Religion* (Düsseldorf 1990) 117–133.
Langdon, John, *Mills in the medieval economy 1300–1540* (Oxford 2004).
Lawrence, Clifford Hugh, *Medieval monasticism,. Forms of religious life in Western Europe in the Middle Ages* (2nd edn., London 1989).
Ledderose, Lothar, *Ten thousand things: Modules and mass production in Chinese art* (Princeton 1998).
Lee, Thomas H.C., 'Sung schools and education before Chu Hsi', in: Wm.Theodore De Bary and John W. Chaffee (eds.), *Neo-Confucian education: The formative stage* (Berkeley 1989) 105–136.
——, *Education in traditional China: a history* (Leiden 2000).
Lefèvre, Wolfgang (ed.), *Picturing machines 1400–1700* (Cambridge Mass. 2004).
—— (ed.), 'Introduction', in: idem (ed.), *Picturing machines 1400–1700* (Cambridge Mass. 2004) 1–10.
Lefèbvre des Noëttes, Richard J.C.E., *L'attelage, le cheval de selle à travers les âges* (Paris 1931).
Lehmann, Hartmut and Guenther Roth (eds.), *Weber's Protestant Ethic. Origins, evidence, context* (Cambridge 1982).
Leidelmeier, Margaret, *Van suikermolen tot grootbedrijf. Technische vernieuwing in de Java-suikerindustrie in de negentiende eeuw* (Amsterdam 1997).
Lemonnier, Pierre (ed.), *Technological choices. The transformations of material cultures since the Neolithic* (London 1993).
Leng, Rainer, 'Social character, pictorial style, and the grammar of technical illustration in craftsmen's manuscripts in the Late Middle Ages', in: Wolfgang Lefèvre (ed.), *Picturing machines 1400–1700* (Cambridge Mass. 2004) 85–111.
Leonard, Jane Kate, *Controlling from afar. The Daoguang Emperor's management of the Grand Canal crisis, 1824–1826* (Ann Arbor 1996).
Lesger, Clé, *Handel in Amsterdam ten tijde van de Opstand. Kooplieden, commerciële expansie en verandering in de ruimtelijke economie van de Nederlanden ca.1550–ca.1630* (Hilversum 2001).

Letts, Malcolm (ed.), *The pilgrimage of Arnold von Harff, knight from Cologne, through Italy, Syria, Egypt, Arabia, Ethiopia, Nubia, Palestine, Turkey, France and Spain, which he accomplished in the years 1496 to 1499* (Hakluyt Society, 2nd series, XCIV) (London 1946).

Leung, Anela Ki Che, 'Elementary education in the Lower Yangtze region in the seventeenth and eighteenth centuries', in: Benjamin A. Elman and Alexander Woodside (eds.), *Education and society in Late Imperial China, 1600–1900* (Berkeley 1994) 381–416.

Lewis, Michael, 'The Hellenistic period', in: Örjan Wikander (ed.), *Handbook of ancient water technology* (Leiden 2000) 631–648.

Leyder, Dirk, 'L'éclosion scolaire. Le développement du réseau des colleges dans les Pays-Bas espagnols et la principauté de Liège', *Paedagogica Historica*, 36 (2000) 1003–1051.

Li, Bozhong, *Agricultural development in Jiangnan 1620–1850* (New York 1988).

——, 'Was there a "fourteenth-century turning point"? Population, land, technology and farm management', in: Paul J. Smith and Richard von Glahn (eds.), *The Song-Yuan-Ming transition in Chinese history* (Cambridge 2003) 134–175.

——, 'Involution and Chinese cotton textile production: Songjiang in the late eighteenth and early nineteenth centuries', in: Giorgio Riello and Prasannan Parthasarathi (eds.), *The spinning world. A global history of cotton textiles, 1200–1850* (Oxford 2009) 387–395.

Li, Lilian, *Fighting famine in North China. State, market and environmental decline, 1690s–1990s* (Stanford 2007).

Lilje, Hanns, *Das technische Zeitalter. Versuch einer biblischen Deutung* (2nd edn., Berlin 1928).

Lindgren, Uta, 'Ordnungsprinzipien in technischen Handschriften und Drucken des Spätmittelalters und der frühen Neuzeit', in: Mathieu Arnoux and Pierre Monnet (eds.), *Le technicien dans la cité en Europe Occidentale, 1250–1650* (Rome 2004) 205–214.

Lishing, Feng and Tong Qingjun, 'Crank-connecting rod mechanism: Its applications in Ancient China and its origin', in: Hong-Sen Yan and Marco Ceccarelli (eds.), *International symposium on History of machines and mechanisms Proceedings of HMM 2008* (s.l. 2009) 235–250.

Liu, Deming, 'Reflections on lack of a patent system throughout China's long history', *Journal of world intellectual property*, 12 (2009) 122–136.

Liu, James T., *China turning inward: Intellectual-political changes in the early twelfth century* (Cambridge Mass. 1988).

Liu, Jerry C.-Y., 'Cultural logics for the regime of useful knowledge during the Ming and early-Qing China c.1400–1700', *History of Technology*, 29 (2009) 29–55.

Liu, Lien Bich, *Immigrants and the industries of London 1500–1700* (Aldershot 2005).

Liu, Xinru, *Silk and religion. An exploration of material life and the thought of people AD 600–1200* (Delhi 1996).

——, *Ancient India and Ancient China. Trade and religious exchange AD 1–600* (1988).

Lohrmann, Dieter, 'Das Maschinenbuch des Konrad Gruter für Erich VII., König von Dänemark (1424)', *Deutsches Archiv für Erforschung des Mittelalters*, 63 (2007) 71–92.

Lombaerde, Piet, 'Introduction', in: Piet Lombaerde (ed.), *Innovation and experiment in the early Baroque in the Spanish Netherlands* (Turnhout 2008) 15–30.

Long, Pamela O., *Openness, secrecy, authorship. Technical arts and the culture of knowledge from Antiquity to the Renaissance* (Baltimore/London 2001).

——, *Technology and society in the medieval centuries. Byzantium, Islam, and the West, 500–1300* (Washington 2003).

——, 'Picturing the machine: Francesco di Giorgio and Leonardo da Vinci in the 1490s', in: Wolfgang Lefèvre (ed.), *Picturing machines 1400–1700* (Cambridge Mass. 2004) 117–141.

——, 'Hydraulic engineering and the study of Antiquity: Rome, 1557–1570', *Renaissance Quarterly*, 61 (2008) 1098–1138.

López Piñero, José María, *Ciencia y técnica en la sociedad española de los siglos XVI y XVII* (Barcelona, 1979).

Lottman, E.B.M., 'De bijdrage van de Amsterdamse weeshuizen aan de bouwkundige opleiding in de achttiende eeuw', *Jaarboek Amstelodamum*, 69 (1977), 140–155.

——, 'De bijdrage van de Rotterdamse en Schiedamse "tekenscholen" aan het bouwkundig onderwijs circa 1750–1850 ', *Rotterdams Jaarboekje* 9th series, 1 (1983) 243–272.

Lourens, Piet and Jan Lucassen, ' "Zunftlandschaften" in den Niederlanden und im benachbarten Deutschland', in: Wilfried Reininghaus (ed.), *Zunftlandschaften in Deutschland und den Niederlanden im Vergleich* (Münster 2000) 11–43.

Love, Harold, *Scribal publication in XVIIth century* (Oxford 1993).

Lucas, Adam, 'The role of the monasteries in the development of medieval milling', in: Steven A. Walton (ed.), *Wind and water in the Middle Ages* (Tempe 2006) 89–127.

——, *Wind, water, work. Ancient and medieval milling technology* (Leiden 2006).

Lucassen, Jan, Piet Lourens and Bert de Munck, 'The distribution of guilds in the Low Countries, 1000–1800', in: Paolo Massa and Angelo Moioli (eds.), *Dalla corporazione al mutuo soccorso. Organizzazione e tutela del lavoro tra XVI e XX secolo* (Milan 2006) 33–56.

Lui, Adam Yuen-Chung, *The Hanlin Academy. Training ground for the ambitious 1644–1850* (Hamden 1981).

Lundgreen, Peter, 'Engineering education in Europe and the U.S.A., 1750–1930. The rise to dominance of school culture and the engineering profession', *Annals of Science*, 47 (1990) 33–75.

Maag, Karin, 'Financing education: The Zurich approach, 1550–1620', in: Beat A. Kümin (ed.), *Reformations old and new. Essays on the socio-economic impact of religious change c.1470–1630* (Aldershot 1996) 203/216.

MacCulloch, Diarmaid, *Reformation. Europe's House divided 1490–1700* (London 2003).

Machlup, Fritz, *The economics of information and human capital* (Princeton 1984).

MacLeod, Christine, *Inventing the Industrial Revolution. The English patent system, 1660–1800* (Cambridge 1988).

Maffioli, Cesare S. and Lodewijk C. Palm (eds.), *Italian scientists in the Low Countries in the XVIIth and XVIIIth centuries* (Amsterdam 1989).

——, *Out of Galileo. The science of waters 1628–1718* (Rotterdam 1994).

Magnusson, Roberta, *Water technology in the Middle Ages. Cities, monasteries, and waterworks after the Roman Empire* (Baltimore 2001).

Malanima, Paolo, *Energia e crescita nell' Éuropa preindustriale* (Rome 1996).

——, 'Energy crisis and growth 1650–1850: The European deviation in a comparative perspective', *Journal of Global History*, 1 (2006) 101–121.

——, *Pre-modern European economy: One thousand years (10th–19th centuries)* (Leiden 2009).

Mandich, Giulio, 'Le privative industriali veneziane, 1450–1550', *Rivista del diritto commerciale*, 34 (1936) I, 518–547.

Mann, Michael, *The sources of social power*, vol. I. *A history of power from the beginning to A.D 1760* (Cambridge 1986).

Maravall, José Antonio, *Antiquos y modernos. Visión de la historia e idea de progreso hasta el Renacimiento* (2nd edn., Madrid 1986).

Marguet, Frédéric Philippe, *Histoire de la longitude à la mer au XVIII*e *siècle en France* (Paris 1917).

Maris, Adriana Johanna, *De reformatie der geestelijke en kerkelijke goederen in Gelderland, in het bijzonder in het Kwartier van Nijmegen* (The Hague 1939).

Mark, Robert, 'Technological innovation in High Gothic architecture', in: Elisabeth Bradford Smith and Michael Wolfe (eds.), *Technology and resource use: cathedrals, mills, mines* (Aldershot 1997) 11–25.

Martin John J., *Venice's hidden enemies. Italian heretics in a Renaissance city* (Berkeley 1993).

Massa, Paolo and Angelo Moioli (eds.), *Dalla corporazione al mutuo soccorso: organizzazione e tutela dal lavoro tra XVI e XX secolo* (Milan 2006).

Mateos Royo, José Antonio, 'State policy, institutional framework and technical monopoly in early modern Spain: invention patents in the Crown of Aragon during the seventeenth century', *History and Technology*, 25 (2009) 147–162.

McClellan III, James E., *Colonialism and science. Saint Domingue in the Old Regime* (2nd edn., Chicago 2010).
McClellan III, James E. and Harold Dorn, *Science and technology in world history. An introduction* (Baltimore 1999).
McConnell, Anita, 'A profitable visit: Luigi Fernando Marsigli's studies, commerce and friendships in Holland, 1722–23', in: Cesare S. Maffioli and Lodewijk C. Palm (eds.), *Italian scientists in the Low Countries in the XVIIth and XVIIIth centuries* (Amsterdam 1989) 189–207.
McCormick, Michael, *Origins of the European economy. Communications and commerce, A.D. 300–900* (Cambridge 2001).
McDermott, Joseph P., *A social history of the Chinese book: Books and literati culture in Late Imperial China* (Hong Kong 2006).
——, 'Book collecting in Jiangxi during the Song dynasty', in: Lucille Chia and Hilde de Weerdt (eds.), *Knowledge and text production in an age of print: China, 900–1400* (Leiden 2011) 63–101.
McGee, David, 'The origins of early modern machine design', in: Wolfgang Lefèvre (ed.), *Picturing machines 1400–1700* (Cambridge Mass. 2004) 53–84.
McManners, John, *Church and society in eighteenth-century France*, 2 vols. (Oxford 1998).
McMullen, David, *State and scholars in T'ang China* (Cambridge 1988).
McPhee, Sarah, *Bernini and the bell towers. Architecture and politics at the Vatican* (New Haven/London 2002).
Meadow, Mark A., 'Merchants and marvels. Hans Jacob Fugger and the origins of the Wunderkammer', in: Pamela Smith and Paula Findlen (eds.), *Merchants and marvels. Commerce, science and art in early modern Europe* (New York/London 2002) 182–202.
Mehl, Lothar, 'Die Anfänge des Navigationsunterricht unter besonderern Berücksichtigung der deutschen Verhältnisse', *Paedagogica Historica*, 8 (1968) 372–441.
Melton, James Van Horn, *Absolutism and eighteenth century origins of compulsory schooling in Prussia and Austria* (Cambridge 1988).
Menzies, Nicholas K., *Forest and land management in Imperial China* (New York 1994).
Merchant, Carolyn, *The death of nature. Women, ecology, and the Scientific Revolution* (San Francisco 1980).
Merton, Robert K., *Science, technology and society in seventeenth-century England* (Bruges 1938).
Meskill, John, *Academies in Ming China. A historical essay* (Tucson 1982).
Métailié, Georges, 'The representation of plants: engravings and paintings', in: Francesca Bray, Vera Dorofeeva-Lichmann and Georges Métailié (eds.), *Graphics and texts in the production of technical knowledge in China. The warp and the weft* (Leiden 2007) 487–520.
Minard, Philippe, *La fortune du colbertisme. État et industrie dans la France des Lumières* (Paris 1998).
Mitcham, Carl and Jim Grote (eds.), *Theology and technology. Essays in Christian analysis and exegesis* (Lanham 1984).
Mitcham, Carl and Jim Grote, 'Aspects of Christian exegesis: hermeneutics, the theological virtues, and technology', in: Carl Mitcham and Jim Grote (eds.), *Theology and technology. Essays in Christian analysis and exegesis* (Lanham 1984) 21–42.
Mitterauer, Michael, *Warum Europa? Mittelalterliche Grundlagen eines Sonderwegs* (Munich 2003).
——, 'Religion und Massenkommunikation. Buchdruck im Kulturvergleich', in: Margarete Grandner and Andrea Komlosy (eds.), *Vom Weltgeist beseelt. Globalgeschichte 1700–1815* (Vienna 2004) 243–262.
Mokyr, Joel, *The lever of riches. Technological creativity and economic progress* (Oxford 1990).
——, *The gifts of Athena. Historical origins of the knowledge economy* (Princeton 2002).
——, *The enlightened economy. An economic history of Britain 1700–1850* (New Haven 2009).

Mol, Hans, 'Aduard 1192–1594. Een korte geschiedenis van het klooster', in: *Een klooster, drie dorpen. Geschiedenis van Aduard, Den Ham en Den Horn* (Bedum 1992) 19–36.
Molà, Luca, *The silk industry of Renaissance Venice* (Baltimore 2000).
Moll-Murata, Christine, 'Chinese guilds from the seventeenth to the twentieth centuries: An overview', *International Review of Social History*, 53 (2008) Supplement 16, 213–247.
——, *States and crafts in the Qing dynasty (1644–1911)* (Tübingen 2008).
Moll-Murata, Christine, Song Jianze and Hans Ulrich Vogel (eds.), *Chinese handicraft regulations of the Qing dynasty* (Munich 2005).
Money, John, 'Teaching in the market place, or "Caesar adsum jam forte: Pompey aderat": the retailing of knowledge in provincial England during the eighteenth century', in: John Brewer and Roy Porter (eds.), *Consumption and the world of goods* (London 1993) 335–377.
Morris-Suzuki, Tessa, 'Concepts of nature and technology in pre-industrial Japan', *East Asian History*, 1 (1991) 81–97.
Mukerji, Chandra, *Impossible engineering. Technology and territoriality on the Canal du Midi* (Princeton 2009).
Mullett, Michael A., *The Catholic Reformation* (London/New York 1999).
Mumford, Lewis, *Technics and civilization* (1934, repr. Orlando 1963).
Mungello, David E., *Curious land: Jesuit accommodation and the origins of Sinology* (Wiesbaden/Stuttgart 1985).
Murray, Peter, *The architecture of the Italian Renaissance* (London 1986).
Musillo, Marco, 'Reconciling two careers: The Jesuit memoir of Giuseppe Castiglione lay brother and Qing imperial painter', *Eighteenth-Century Studies*, 42 (2008) 45–59.
Mijnhardt, Wijnand W., *Tot heil van 't menschdom. Culturele genootschappen in Nederland, 1750–1815* (Amsterdam 1988).
——, 'The Dutch Enlightenment: Humanism, nationalism and decline', in: Margaret C. Jacob and Wijnand W. Mijnhardt (eds.), *The Dutch Republic in the eighteenth century. Decline, enlightenment and revolution* (Ithaca 1992) 197–223.
——, 'The construction of silence' in: Wiep van Bunge (ed.), *The early Enlightenment in the Dutch Republic, 1650–1750* (Leiden 2003) 231–262.
Nakayama, Shigeru, *Academic and scientific traditions in China, Japan and the West* (Tokyo 1984).
Nalle, Sarah, 'Literacy and culture in early modern Castile', *Past and Present*, number 125 (1989) 65–96.
Naquin, Susan, 'Temples, technology, and material culture in Shouzhou, Anhui', in: Dagmar Schäfer (ed.), *Cultures of knowledge. Technology in Chinese history* (Leiden 2012) 185–207.
Naquin, Susan and Chün-fang Yü, 'Introduction', in: idem (eds.), *Pilgrims and sacred sites in China* (Berkeley 1992) 1–38.
Naquin, Susan and Evelyn Rawski, *Chinese society in the eighteenth century* (New Haven 1987).
Navarro-Brotóns, Víctor, 'The teaching of the mathematical disciplines in sixteenth-century Spain', *Science and Education*, 15 (2006) 209–233.
Navarro-Brotóns, Víctor and William Eamon, 'Spain and the Scientific Revolution: Historiographical questions and conjectures', in: Víctor Navarro-Brotóns and William Eamon (eds.), *Más allá de la Leyenda Negra. España y la Revolución Científica/Beyond the Black Legend: Spain and the Scientific Revolution* (Valencia 2007) 27–38.
Needham, Joseph, Wang Ling and Derek de Solla Price, *Heavenly clockwork. The great astronomical clocks of medieval China* (Cambridge 1960).
——, *The Grand Titration. Science and society in East and West* (London 1969).
——, *Science in traditional China* (Cambridge Mass. 1981).
——, *Science and civilisation in China*, vol. I (Cambridge 1954).
——, et al., *Science and civilisation in China*, vol. II, *History of scientific thought* (Cambridge 1956), vol. IV *Physics and physical technology*, part 2 *Mechanical engineering* (Cambridge

1965), part 3, *Civil engineering and nautics* (Cambridge 1971), vol. V *Chemistry and chemical technology*, part 9, *Textile technology, spinning and reeling* (Cambridge 1988), part 11, *Ferrous metallurgy* (Cambridge 2008), part 12, *Ceramic technology* (Cambridge 2004), part 13, *Mining* (Cambridge 1999), vol. VI *Biology and biological technology*, part 1, *Botany* (Cambridge 1986), part 3, *Agro-industries: Sugarcane technology* (Cambridge 1996), part 5, *Fermentations and food science* (Cambridge 2000).

Nelson, Benjamin, 'Science, civilisations, "East and West"', *Boston Studies in the Philosophy of Science*, 11 (1974) 445–493.

Neugebauer, Wolfgang, 'Niedere Schulen und Realschulen', in: Christa Berg et al. (eds.), *Handbuch der deutsche Bildungsgeschichte*, bd.II, ed. by Notker Hammerstein and Ulrich Hammerstein (Munich 2005) 213–261.

Newman, Martha G., *The boundaries of charity. Cistercian culture and ecclesiastical reform 1098–1180* (Stanford 1996).

Nierop, Henk van, *Van ridders tot regenten. De Hollandse adel in de zestiende en de eerste helft van de zeventiende eeuw* (Amsterdam 1984).

Nierop, Leonie van, 'Gegevens over de nijverheid van Amsterdam bijeengelezen uit de advertenties in de Amsterdamsche Courant', *Jaarboek Amstelodamum*, 27 (1930), 264–311, 28 (1931) 95–182.

Nieuw Nederlandsch Biografisch Woordenboek, vol. IX (Leiden 1937).

Noble, David, *The religion of technology. The divinity of man and the spirit of invention* (New York 1998).

Norborg, Lars-Anne, *Storföretaget Vadstena kloster. Studier in senmedeltida godspolitik och ekonimiförvaltning* (Lund 1958).

Norman, Herbert and H. John, *The organ today* (New York 1966).

North, Douglass C. and Robert Paul Thomas, *The rise of the Western world. A new economic history* (Cambridge 1973).

Nugent, Thomas, *The Grand Tour containing an exact description of the most of the cities, towns and remarkable places of Europe* (London 1749).

Numbers, Ronald L. (ed.), *Galileo goes to jail and other myths about science and religion* (Cambridge 2009).

Nuñez, Clara Eugenia, 'Literacy, schooling and economic modernization', *Paedagogica Historica*, 39 (2003) 535–558.

Nylan, Michael, *The five "Confucian" classics* (New Haven 2001).

Oakley, Francis, *Natural law, laws of nature, natural rights. Continuity and discontinuity in the history of ideas* (New York 2005).

O'Brien, Patrick K., 'The Needham Question updated: A historiographical survey and elaboration', *History of Technology*, 29 (2009) 7–28.

——, 'Ten years of debate on the origins of the Great Divergence', *Reviews in History*, 1008 (www.history.ac.uk/reviews/review/1008).

O'Malley, John W. (ed.), *The Jesuits. Culture, science and the arts, 1540–1773* (Toronto 1999).

——, *Trent and all that. Renaming Catholicism in the Early Modern Era* (Cambridge Mass. 2000).

——, et al. (ed.), *The Jesuits II. Culture, science and the arts, 1540–1773* (Toronto 2006).

Orme, Nicholas, *Medieval schools from Roman Britain to Renaissance England* (New Haven 2006).

Osborne, Anne Rankin, *Barren mountains, raging rivers. The ecological and social effects of changing land use on the Lower Yangzi periphery in Late Imperial China* (Ph.D. Thesis, University of Columbia 1989).

Overton, Mark, *Agricultural revolution in England. The transformation of the agrarian economy 1500–1850* (Cambridge 1996).

Ovitt, George, *The restoration of perfection. Labor and technology in medieval culture* (New Brunswick 1987).

Pagani, Catherine, *"Eastern magnificence and European ingenuity". Clocks in Late Imperial China* (Ann Arbor 2001).

Pallach, Ulrich-Christian, 'Fonctions de la mobilité artisanale et ouvrière- compagnons, ouvriers et manufacturiers en France et aux Allemagnes (17e–19e siècles)', *Francia*, 11 (1983) 365–406.

Pamuk, Sevket, 'The Black Death and the origings of the "Great Divergence" across Europe, 1300–1600', *European Review of Economic History*, 11 (2007) 389–317.

Parthasarathi, Prasannan, 'Rethinking wages and competitiveness in the eighteenth century: Britain and South India', *Past and Present*, number 158 (1998) 79–109.

——, 'The Great Divergence', *Past and Present*, 176 (2002) 275–293.

——, *Why Europe grew rich and Asia did not. Global economic divergence, 1600–1850* (Cambridge 2011).

Partridge, Loren, *The Renaissance in Rome 1400–1600* (London 1996).

Passmore, John, *Man's responsibility for nature. Ecological problems and Western traditions* (New York 1974).

Peiffer, Jeanne, 'Projections embodied in technical drawings: Dürer and his followers', in: Wolfgang Lefèvre (ed.) *Picturing machines 1400–1700* (Cambridge Mass. 2004) 245–275.

Perdue, Peter C., 'Water control in the Dongting Lake region during the Ming and Qing periods', *Journal of Asian Studies*, 41 (1982) 747–765.

——, 'How different was China? Or, bringing the army back in: coercion and ecology in the comparative sociology of Europe and China', in: Rolf Sieferle and Helga Breuninger (eds.), *Agriculture, population and economic development in China and Europe* (Stuttgart 2003) 311–330.

Persson, Karl-Gunnar, *Pre-industrial economic growth. Social organization and technological progress in Europe* (New York 1988).

——, *An economic history of Europe. Knowledge, institutions and growth, 600 to the present* (Cambridge 2010).

Peters, Frank E., *Jerusalem. The Holy City in the eyes of chroniclers, visitors, pilgrims, and prophets from the days of Abraham to the beginnings of modern times* (Princeton 1985).

——, *Jerusalem and Mecca. The typology of the Holy City in the Near East* (New York 1986).

Petrucci, Armando, *Writers and readers in medieval Italy. Studies in the history of written culture* (New Haven 1995).

Pettegree, Andrew (ed.), *The early reformation in Europe* (Cambridge 1992).

Philipp, Klaus Jan, '"Eyn huys in manieren van eynre kirchen". Werkmeister, Parliere, Steinlieferanten, Zimmermeister unde Bauorganisation in den Niederlanden vom 14. bis zum 16 Jahrhundert', *Wallraf-Richartz Jahrbuch*, 50 (1989) 69–113.

Po-Chia Hsia, Ronnie, *The world of Catholic Renewal 1540–1770* (Cambridge 1998).

Po-Chia Hsia, Ronnie and Henk van Nierop (eds.), *Calvinism and religious toleration in the Dutch Golden Age* (Cambridge 2002).

Poelwijk, Arjen, *'In dienste vant suyckerbacken', De Amsterdamse suikernijverheid en haar ondernemers, 1580–1630* (Hilversum 2003).

Poitrineau, Abel, 'L'Organisation du travail et ses dysfonctionnements en France du XVIe et XVIe siècle', *Journal of Medieval and Renaissance Studies*, 25 (1995) 107–127.

Pomeranz, Kenneth, *The Great Divergence. Europe, China, and the making of the modern world economy* (Princeton 2000).

Popplow, Marcus, *Neu, nützlich und erfindungsreich. Die Idealisierung von Technik in der frühen Neuzeit* (Münster 1998).

——, *Models of machines: A 'missing link' between early modern engineering and mechanics?* (Preprint nr.225 Max Planck-Institut für Wissenschaftsgeschichte, Berlin 2002).

——, 'Why draw pictures of machines? The social contexts of early modern machine drawings', in: Wolfgang Lefèvre (ed.), *Picturing machines 1400–1700* (Cambridge Mass. 2004) 17–48.

——, *Technik im Mittelalter* (Munich 2010).

Porter, Roy (ed.), *The Cambridge History of Science*, vol. 4 *Eighteenth-century science* (Cambridge 2003).

Portuondo, María M., *Secret science. Spanish cosmography and the New World* (Chicago/London 2009).

Pot, Johan Hendrik Jacob van der, *Die Bewertung des technisches Fortschritts* (Assen 1985).

Prak, Maarten, 'The politics of intolerance: citizenship and religion in the Dutch Republic (seventeenth to eighteenth centuries)', in: Ronnie Po-Chia Hsia and Henk van Nierop (eds.), *Calvinism and religious toleration in the Dutch Golden Age* (Cambridge 2002) 159–175.

Prak, Maarten, Catharina Lis, Jan Lucassen and Hugo Soly (eds.), *Craft guilds in the early modern Low Countries* (Aldershot 2006).

Prévot, André, *L'enseignement technique chez les Frères des Écoles Chrétiennes au XVII*[e] *et au XIX*[e] *siècles* (Paris 1964).

Prip-Møller, Johannes, *Chinese Buddhist monasteries: their plan and its function as a setting for Buddhist monastic life* (Copenhagen 1937, 2nd edn., Hong Kong 1967).

Queen, Sara A., 'Inventories of the past: rethink the "school" affiliation of the Huainanzi', *Asia Major*, series 3, 14:1 (2001) 51–72.

Raj, Kapil, *Relocating science. Circulation and the construction of knowledge in South Asia and Europe 1650–1900* (Basingstoke 2006).

Rapp, Richard T., *Industry and the economic decline in seventeenth-century Venice* (Cambridge Mass. 1976).

Rawski, Evelyn, *Education and popular literacy in Ch'ing China* (Ann Arbor 1979).

Reis, Jaime, 'Economic growth, human capital formation and consumption in Western Europe before 1800', in: Robert C. Allen, Tommy Bengtsson and Martin Dribe (eds.), *Living standards in the past. New perspectives on well-being in Asia and Europe* (Oxford 2005) 195–225.

Reischauer, Edwin O., *Ennin's diary. The record of a pilgrimage to China in search of the law* (New York 1955).

Reith, Reinhold, 'Arbeitsmigration und Technologietransfer in der Habsburgermonarchie in der zweiten Hälfte des 18. Jahrhunderts', in: Ulrich Troitzsch (ed.), *'Nützliche Künste'. Kultur- und Sozialgeschichte der Technik im 18. Jahrhundert* (Münster 1999) 51–66.

——, 'Circulation of skilled labour in late medieval and early modern Central Europe', in: Larry Epstein and Maarten Prak (eds.), *Guilds, innovation and the European economy, 1400–1800* (Cambridge 2008) 114–142.

Reynolds, Terry S., *Stronger than a hundred men. A history of the vertical water wheel* (Baltimore 1983).

Ricci, Matteo and Nicholas Trigault, *Histoire de l'expedition chrétienne au royaume de la Chine 1582–1610*, ed. Georges Bessière (1978, original edition in Latin, Augsburg 1615).

Richardson, Gary, 'Christianity and craft guilds in Late Medieval England. A rational choice analysis', *Rationality and Society*, 17 (2005) 139–189.

Ridder-Symoens, Hilde de, 'La sécularisation de l'enseignement aux anciens Pays-Bas au moyen âge et à la Renaissance', in: Jean-Marie Duvosquel and Erik Thoen (eds.), *Peasants and Townsmen in medieval Europe* (Gent 1995) 721–739.

Roelfsema, Ena Hebelina, *De klooster- en proosdijgoederen in de provincie Groningen* (Groningen 1928).

Roetz, Heiner, *Mensch und Natur im alten China* (Frankfurt am Main 1984).

——, 'On nature and culture in Zhou China', in: Hans Ulrich Vogel and Günter Dux (eds.), *Concepts of nature. A Chinese-European cross-cultural perspective* (Leiden 2010) 198–219.

Roorda, Daniel Jeen, 'Contrasting and converging patterns: Relations between church and state in Western Europe, 1660–1715', in: Alastair C. Duke and Coen A. Tamse (eds.), *Britain and the Netherlands* VII (The Hague 1981) 134–153.

Rosenberg, David Lee (ed.), *Les protestants amiènois au milieu du XVI*[e] *siècle* (Amiens 1994).

Rosenthal, Jean-Laurent and Roy Bin Wong, *Before and beyond divergence. The politics of economic change in China and Europe* (Cambridge Mass. 2011).

Ross, Andrew C., *A vision betrayed. The Jesuits in Japan and China, 1542–1742* (Edinburgh 1994).

Rowe, William T., *Hankow. Commerce and society in a Chinese city, 1796–1889* (Stanford 1984).
——, *China's last empire. The great Qing* (Cambridge Mass. 2009).
——, 'Political, social and economic factors affecting transmission of technical knowledge in early modern China', in: Dagmar Schäfer (ed.), *Cultures of knowledge. Technology in Chinese history* (Leiden 2012) 25–44.
Roy, Tirthankar, 'Knowledge and divergence from the perspective of early modern India', *Journal of Global History*, 3 (2008) 361–387.
Rubiés, Joan-Pau, 'Travel writing as a genre: facts, fictions and the invention of a scientific discourse in early modern Europe', in: idem, *Travellers and cosmographers. Studies in the history of early modern travel and ethnology* (Aldershot 2007), I (5–35).
——, 'Travel writing and ethnography', in: idem, *Travellers and cosmographers. Studies in the history of early modern travel and ethnology* (Aldershot 2007), IV (1–39).
——, 'Instructions for travellers: teaching the eye to see', in: idem, *Travellers and cosmographers. Studies in the history of early modern travel and ethnology* (Aldershot 2007), III (139–190).
Ruitenbeek, Klaas *Carpentry and building in Late Imperial China. A study of the fifteenth-century carpenter's manual Lu Ban jing* (Leiden 1993).
Russo, François, 'L'hydrographie en France aux XVIIe et XVIIIe siècles. Écoles et ouvrages d'enseignement', in: René Taton (ed.), *Enseignement et diffusion des sciences en France au XVIIIe siècle* (Paris 1964) 419–440.
Sambursky, Samuel, *The physical world of the Greeks* (London 1956).
Sauzet, Robert, *Contre-Réforme et réforme catholique en Bas-Languedoc. Le diocèse de Nîmes au XVIIe siècle* (Lille 1978).
Sawday, Jonathan, *Engines of the imagination. Renaissance culture and the rise of the machine* (New York 2007).
Schäfer, Dagmar, 'The congruence of knowledge and action: The Tiangong kaiwu and its author Song Yingxing', in: Christine Moll-Murata, Song Jianze and Hans Ulrich Vogel (eds.), *Chinese handicraft regulations of the Qing dynasty* (Munich 2005) 35–60.
——, *The crafting of 10,000 things. Knowledge and technology in seventeenth-century China* (Chicago 2011).
——, 'Silken strands: Making technology work in China', in: Dagmar Schäfer (ed.), *Cultures of knowledge. Technology in Chinese history* (Leiden 2012) 45–73.
—— (ed.), *Cultures of knowledge. Technology in Chinese history* (Leiden 2012).
Schaik, Peter van, *Christiaan Brunings 1736–1805. Waterstaat in opkomst* (Zutphen 1984).
Scharpf, Fritz W. 'Games real actors could play: positive and negative coordination in embedded negotiations', *Journal of Theoretical Politics* 6 (1994) 27–53.
Scheuring, Hans Lothar, *Die Drachenfluß-Werft von Nanking. Das Lung-chiang ch'uan-ch'ang chich, eine Ming-zeitliche Quelle zur Geschichte des chinesischen Schiffbaus* (Frankfurt and Main 1987).
Schiebinger, Londa, *The mind has no sex. Women in the origins of modern science* (Cambridge Mass. 1989).
Schiebinger, Londa, *Nature's body. Gender in the making of modern science* (Boston 1993).
Schilling, Heinz, 'Confessional migration and social change. The case of the Dutch refugees of the sixteenth century', in: Paul Klep and Eddy van Cauwenberghe (eds.), *Entrepreneurship and the transformation of the economy (10th–20th centuries)* (Leuven 1994) 321–333.
Schmidt-Glintzer, Helwig, 'On the relationship between man and nature in China', in: Hans Ulrich Vogel and Günter Dux (eds.), *Concepts of nature. A Chinese-European cross-cultural perspective* (Leiden 2010) 526–542.
Schneider, Richard, 'Un moine Indien au Wou-t'ai chan. Relation d'un pélerinage', *Cahiers d'Extrême Asie*, 3 (1987) 27–39.
Schott, Kaspar, *Mechanica hydraulico-pneumatica* (Würzburg 1657).
——, *Technica curiosa, sive mirabilis artis* (Würzburg 1664).

Schulz-Forberg, Hagen, 'European travel and travel writing. Cultural practice and the idea of Europe', in: idem (ed.), *Unravelling civilisation. European travel and travel writing* (Brussels 2005) 13–40.
Scott, Robert A., *The Gothic enterprise. A guide to understanding the medieval cathedral* (Berkeley 2003).
Scott, Rosemary E. (ed)., *The porcelains of Jingdezhen* (London 1993).
Scoville, Warren C. *The persecution of the Huguenots and French economic development, 1680–1720* (Berkeley 1960).
——, 'Minority migration and the diffusion of technology', *Journal of Economic History*, 11 (1951) 347–360.
——, 'The Huguenots and the diffusion of technology', *Journal of Political Economy*, 50 (1952) 294–311, 392–411.
Secord, James A., 'Knowledge in transit', *Isis* 95 (2004) 654–672.
Seifert, Arno, 'Das höhere Schulwesen. Universitäten und Gymnasien', in: Christa Berg et al. (eds.), *Handbuch der deutsche Bildungsgeschichte*, bd.II, ed. by Notker Hammerstein (Munich 1996) 197–374.
Semedo, Alvaro, *Relazione della grande monarchia della Cina* (Rome 1643).
Shapiro, Barbara J., *A culture of fact. England, 1550–1750* (Ithaca 2000).
Shelby, Lon R., 'The role of the master mason in medieval English building', *Speculum* 39 (1964) 387–403.
——, 'The geometrical knowledge of mediaeval master masons' *Speculum*, 47 (1972), 395–421.
Simpson, James, *Burning to read. English fundamentalists and its Reformation opponents* (Cambridge 2007).
Siebert, Martina, 'Making technology history', in: Dagmar Schäfer (ed.), *Cultures of knowledge. Technology in Chinese history* (Leiden 2012) 253–281.
Sieferle, Rolf Peter, *Das unterirdische Walt. Energiekrise und Industrielle Revolution* (Munich 1982).
——, *Der Europäische Sonderweg. Ursachen und Faktoren* (Stuttgart 2003).
——, 'Why did industrialization start in Europe (and not in China)', in: Rolf Sieferle and Helga Breuninger (eds.), *Agriculture, population and economic development in China and Europe* (Stuttgart 2003) 7–89.
Silk, Jonathan A., 'Marginal notes on a study of Buddhism, economy and society in China', *Journal of the International Association of Buddhist Studies*, 22(2) (1999) 359–396.
——, *Managing monks. Administrators and administrative roles in Indian Buddhist monasticism* (Oxford 2008).
Sivin, Nathan, 'On the word "Taoist" as a source of perplexity: with special reference to the relations of science and religion in traditional China', *History of Religions*, 17 (1978) 303–330.
——, 'Chinesische Wissenschaft. Ein Vergleich der Ansätze von Max Weber und Joseph Needham', in: Wolfgang Schluchter (ed.), *Max Webers Studie über Konfuzianismus und Taoismus. Interpretation und Kritik* (Frankfurt 1983), 342–362.
Sivin, Nathan and Geoffrey Lloyd, *The way and the word. Science and medicine in early China and Greece* (New Haven 2002).
Skinner, G. William (ed.), *The city in Late Imperial China* (Stanford 1977).
——, 'Presidential address: The structure of Chinese history', *Journal of Asian Studies*, 44 no. 2 (February 1985) 271–292.
Smil, Vaclav, *Energy in world history* (San Francisco/Oxford 1994).
Smith, Elisabeth Bradford and Michael Wolfe (eds.), *Technology and resource use: cathedrals, mills, mines* (Aldershot 1997).
Smith, Pamela H. and Paula Findlen, *Merchants and marvels. Commerce, science, and art in early modern Europe* (New York 2002).
——, *The body of the artisan. Art and science in the Scientific Revolution* (Chicago 2004).

Sonkajärvi, Hanna, 'From German-speaking Catholics to French carpenters: Strasbourg guilds and the role of confessional boundaries in the inclusion and exclusion of foreigners in the eighteenth century', *Urban History*, 35 (2008) 202–215.
Spence, Jonathan, *The search for modern China* (New York 1990).
Spufford, Margaret, 'Literacy, trade and religion in the commercial centres of Europe', in: Karel Davids and Jan Lucassen (eds.), *A miracle mirrored. The Dutch Republic in European perspective* (Cambridge 1995) 229–283.
Squatriti, Paolo, *Water and society in early medieval Italy, AD 400–1000* (Cambridge 1998).
Stagl, Justin, *Geschichte der Neugier. Die Kunst des Reisens 1550–1800* (Vienna 2002).
Steidl, Annemarie, *Auf nach Wien! Die Mobilität des mitteleuropäischen Handwerks im 18. und 19. Jahrhundert am Beispiel der Haupt- und Residenzstadt* (Munich 2003).
Stephens, William B., 'Literacy in England, Scotland, and Wales, 1500–1900', *History of Education Quarterly*, 30 (1990), 545–571.
Stewart, Larry, *The rise of public science. Rhetoric, technology and natural philosophy in Newtonian Britain, 1660–1750* (Cambridge 1992).
——, 'Experimental spaces and the knowledge economy', *History of Science*, 45 (2007) 155–177.
Stöcklein, Ansgar, *Leitbilder der Technik. Biblische Tradition und technische Fortschritt* (München 1969).
Stöcklein, Ansgar and Mohammed Rassem (eds.), *Technik und Religion* (Düsseldorf 1990).
Stoll, Mark, *Protestantism, capitalism and nature in America* (Albuquerque 1997).
Stone, Lawrence, 'The educational revolution in England, 1560–1640', *Past and Present* 28 (1964) 41–80.
Storper, Michael and Richard Walker, *The capitalist imperative. Territory, technology and industrial growth* (Oxford 1989).
Strassberg, Richard E., *The world of K'ung Shang-Jen. A man of letters in Early Ch'ing China* (New York 1983).
Stutzer, Dietmar, *Klöster als Arbeitgeber um 1800. Die bayerischen Klöster als Unternehmenseinheiten und ihre Sozialsysteme zur Zeit der Säkularisation um 1800* (Göttingen 1986).
Stuijvenberg, Johannes H. van, 'The Weber Thesis. An attempt at interpretation', *Acta Historiae Neerlandicae*, VIII (1975) 50–66.
Swan, Claudia, *Art, science and witchcraft in early modern Holland. Jacques de Gheyn II (1565–1625)* (Cambridge 2005).
Swatos, William H.jr. and Lutz Kaelber (eds.), *The Protestant Ethic turns 100. Essays on the Centenary of the Weber Thesis* (Boulder/London 2005).
Swetlitz, Marc, 'A Jewish commentary on the religious origins of technological civilization', *Research in Philosophy and Technology*, 6 (1983) 197–204.
Swetz, Frank J., *Capitalism and arithmetic. The new math of the fifteenth century* (La Salle Ill. 1987).
Swigchem, Cornelis Albertus van, *Abraham van der Hart 1747–1820. Architect, stadsbouwmeester van Amsterdam* (Amsterdam 1965).
Tann, Jennifer, 'The textile millwright in the Early Industrial Revolution', *Textile History*, 5 (1974) 80–89.
Taylor, Charles, *A secular age* (Cambridge Mass. 2007).
Taylor, Rodney Leon, *The religious dimensions of Confucianism* (Albany 1990).
Teiser, Stephen A., 'The growth of purgatory', in: Patricia Ebrey and Peter Gregory (eds.), *Religion and society in T'ang and Sung China* (Honolulu 1993) 115–145.
Temple, Katherine, 'Doubts concerning the religious origins of technological civilization', *Research in Philosophy and Technology*, 6 (1983) 189–197.
Ter Haar, Bas, 'Buddhist-inspired options: Aspects of lay religious life in the lower Yangzi from 1100 until 1340', *T'oung Pao*, 87 (2001) 92–152.
Theunissen, Bert, *Nut en nog eens nut. Wetenschapsbeelden van Nederlandse natuuronderzoekers 1800–1900* (Hilversum 2000).

Thomas, Keith, *Religion and the decline of magic* (London 1971).

Thorner, Isidor, 'Ascetic Protestantism and the development of science and technology', *American Sociological Review*, 58 (1952) 25–33.

Thijs, Alfons K.L, 'Religion and social structure. Religious rituals in pre-industrial trade associations in the Low Countries', in: Maarten Prak, Catharina Lis, Jan Lucassen and Hugo Soly (eds.), *Craft guilds in the early modern Low Countries* (Aldershot 2006) 157–173.

Tichane, Robert, *Ching-te-chen: Views of a porcelain city* (New York 1983).

Tikoff, Valentina K., 'Saint Elmo's orphans: Navigation education and training at the Royal School of San Telmo in Seville during the eighteenth century', *International Journal of Maritime History*, 20 (2008) 1–32.

Tresse, René, 'Le Conservatoire des Arts et Métiers et la Société d'Encouragement pour l'Industrie nationale au début du XIXe siècle', *Revue d'histoire des sciences et de leurs applications*, 5 (1952) 246–264.

Tri, Christine Nguyen and Catherine Despeux (eds.), *Éducation et instruction en Chine:* I.: *l'éducation élémentaire* (Éditions Peeters: Paris/Louvain, 2003); II: *Les formations spécialisées* (Éditions Peeters: Paris/Louvain, 2003); III *Aux marges de l'orthodoxie* (Éditions Peeters: Paris/Louvain, 2004).

Troitzsch, Ulrich, *Ansätze technologischer Denkens bei der Kameralisten des 17. und 18. Jahrhunderts* (Berlin 1966).

Troitzsch, Ulrich and Gabrielle Wohlauf (eds.), *Technik-Geschichte. Historische Ansätze und neuere Beiträge* (Frankfurt 1980).

Tschopp, Silvia Serena, 'Popularisierung gelehrten Wissens im 18. Jahrhundert. Institutionen und Medien', in: Richard von Dülmen and Sina Rauschenbach (eds.), *Macht des Wissens. Die Entstehung der modernen Wissensgesellschaft* (Cologne/Weimar/Vienna 2004) 469–489.

Tunzelmann, Nick von, *Technology and industrial progress. The foundations of economic growth* (Cheltenham 1995).

Turnbull, David, 'The ad-hoc collective works of building Gothic cathedrals with templates, string and geometry', *Science, Technology and Human Values*, 18 (1993) 315–340.

Unger, Richard W., 'Energy sources for the Dutch Golden Age: peat, wind and coal', *Research in Economic History*, 33 (1984) 221–253.

Ven, Gerard P. van de, *Aan de wieg van Rijkswaterstaat. Wordingsgeschiedenis van het Pannerdens Kanaal* (Zutphen 1976).

Vermeer, Edward B., 'P'an Chi-hsün's solutions for the Yellow River problems of the late sixteenth century', *T'oung Pao*, 73 (1987) 33–67.

—— (ed.), *Development and decline of Fukien province in the seventeenth and eighteenth centuries* (Leiden 1990).

Vermij, Rienk H., *Secularisering en natuurwetenschap in de zeventiende en achttiende eeuw* (Utrecht 1991).

Viñao Frago, Antonio, 'Alfabetización, lectura y escritura en el Antiguo Régimen (siglos XVI–XVIII)', in: Agustín Escolano (ed.), *Leer y escribir en España. Doscientos años de alfabetización* (Madrid 1992) 45–68.

Vissière, Isabelle and Jean-Louis (eds.), *Lettres édifiantes et curieuses de Chine par des missionaires jésuites 1701–1776* (Paris 1979).

Vogel, Hans Ulrich *Untersuchungen über die Salzgeschichte von Sichuan (311 v. Chr. – 1911)* (Stuttgart 1990).

——, 'The transfer of mining and smelting technology between Asia and Europe in the sixteenth to early nineteeth centuries', *Journal of the Japan-Netherlands Institute*, III (1991) 74–101.

——, 'Technische Entwicklungen und ihre Grenzen im Salzgewerbe von Furong, Sichuan, 17.–19.Jahrhundert', *Orientierungen*, 2 (1991) 89–119.

——, 'Kupfererzeugung und – handel in China und Europa. Mitte des 8. bis Mitte des 19. Jahrhunderts: Eine vergeleichende Studie', *Bochumer Jahrbuch zur Ostasienforschung*, 15 (1991) 1–57.

——, 'The mining industry in traditional China: Intra- and intercultural comparisons', in: Helga Nowotny (ed.), *Cultures of technology and the quest for innovation* (New York 2006) 167–188.

——, '"That which soaks and descends becomes salty": The concept of nature in traditional Chinese salt production', in: Hans Ulrich Vogel and Günter Dux (eds.), *Concepts of nature. A Chinese-European cross-cultural perspective* (Leiden 2010) 469–515.

Vogel, Hans Ulrich and Günter Dux (eds.), *Concepts of nature. A Chinese-European cross-cultural perspective* (Leiden 2010).

Voiret, Jean Pierre, 'Religionen und Technik in China', in: Ansgar Stöcklein, and Mohammed Rassem (eds.), *Technik und Religion* (Düsseldorf 1990) 85–101.

Vovelle, Michel, *Piété baroque et dechristianisation en Provence au XVII*[e] *siècle* (Paris 1973).

Vries, Peer, 'Are coal and colonies really crucial? Kenneth Pomeranz and the Great Divergence', *Journal of World History*, 12 (2001) 407–446.

——, *Via Peking back to Manchester. Britain, the Industrial Revolution, and China* (Leiden 2003).

Vries, Jan de, *The Dutch rural economy in the Golden Age, 1500–1700* (New Haven/London 1974).

Wade, Geoff, 'The Zheng He voyages: A reassessment', *Journal of the Malaysia Branch of the Royal Asiatic Society*, 78 (2005) 37–58.

Wagner, Donald B., *Iron and steel in ancient China* (Leiden 1993).

——, *The traditional Chinese iron industry and its modern fate* (Richmond 1997).

——, 'Song Yingxing's illustrations of iron production', in: Francesca Bray, Vera Dorofeeva-Lichmann and Georges Métailié (eds.), *Graphics and texts in the production of technical knowledge in China. The warp and the weft* (Leiden 2007) 615–632.

Wake, Christopher, 'The myth of Zheng He's great treasure ships', *International Journal of Maritime History*, 16 (2004) 59–75.

Waley-Cohen, Joanna, *The sextants of Beijing. Global currents in Chinese history* (New York 1999).

Wall, Ernestine van der, 'The religious contexts of the early Dutch Enlightenment. Moral religion and society', in: Wiep van Bunge (ed.), *The Early Enlightenment in the Dutch Republic, 1650–1750* (Leiden 2003) 39–62.

Wallace, Peter C., *Communities and conflict in early modern Colmar 1575–1730* (Atlantic Highlands 1995).

Wallis, Patrick, 'Apprenticeship and training in pre-modern England', *Journal of Economic History*, 68 (2008) 832–861.

Watson, Andrew M., *Agricultural innovation in the early Islamic world. The diffusion of crops and farming techniques, 700–1100* (Cambridge 1983).

Webb, Diana, *Pilgrimage in medieval England* (London 2000).

——, *Medieval European pilgrimage c.700–c.1500* (Basingstoke 2002).

Weber, Max, 'Die Wirtschafsethik der Weltreligionen', in: *Gesammelte Aufsätze zur Religionssoziologie*, I (Tübingen 1920, repr. 1988).

——, *Religion und Gesellschaft. Gesammelte Aufsätze zur Religionssoziologie* (Frankfurt am Main 2000).

——, *Wirtschaft und Gesellschaft. Grundriss der verstehende Soziologie* (Tübingen 1972).

Webster, Charles, *The Great Instauration. Science, medicine and reform, 1626–1660* (London 1975).

Weiss, Eberhard, *Die Säkularisation der bayerischen Klöster 1802/03* (Munich 1983).

Weller, Robert P., *Discovering nature. Globalization and environmental culture in China and Taiwan* (Cambridge 2006).

White, Kenneth D., *Greek and Roman technology* (London 1984).

White jr., Lynn, 'The historical roots of our ecological crisis', *Science*, 155 (10 March 1967) 1203–1207.

——, *Machina ex deo. Essays in the dynamism of Western culture* (Cambridge, Mass. 1968).

——, *Medieval religion and technology. Collected essays* (Berkeley 1978) 217–253.

——, 'The Crusades and the technological thrust of the West', in: idem, *Medieval religion and technology. Collected essays* (Berkeley 1978) 277–296.
——, 'Cultural climates and technological advances in the Middle Ages', in: idem, *Medieval religion and technology. Collected essays* (Berkeley 1978), 217–253.
——, 'Western technology', in: Joseph R. Strayer (ed.) *Dictionary of the Middle Ages*, vol. XI (New York 1988) 650–664.
Whitney, Elspeth, 'Lynn White, ecotheology and history', *Environmental Ethics*, 15 (1993) 151–160.
Wikander, Örjan 'The water-mill' in: Örjan Wikander (ed.), *Handbook of ancient water technology* (Leiden 2000) 371–400.
——, 'Industrial applications of water-power', in Örjan Wikander (ed.), *Handbook of ancient water technology* (Leiden 2000) 401–410.
—— 'The Roman empire', in: Örjan Wikander (ed.), *Handbook of ancient water technology* (Leiden 2000) 649–660.
Will, Étienne, 'Un cycle hydraulique en Chine: la province du Hubei du XVIe au XIXe siècles', *Bulletin de l'École française d'Extrème Orient*, 68 (1980) 261–287.
——, 'Développement quantitatif et développement qualitatif en Chine à l' époque impériale', *Annales: Histoire et Science Sociales*, 49 (1994) 863–902.
——, 'L'apprentissage du metier de fonctionnaire à la fin de la période impériale', in: Christine Nguyen Tri and Catherine Despeux (eds.), *Éducation et instruction en Chine*: II: *Les formations spécialisées* (Éditions Peeters: Paris/Louvain, 2003) 7–47.
Williams, Peter, *The European organ 1450–1850* (London 1966).
——, *The organ in western culture, 750–1250* (Cambridge 1993).
Wilson, Andrew, 'Machines, power and the ancient economy', *The Journal of Roman Studies*, 92 (2002) 1–32.
Wilson, Catherine, 'From limits to laws. The construction of the nomological image of nature in early modern philosophy', in: Lorraine Daston and Michael Stolleis (eds.), *Natural law and laws of nature in early modern Europe. Jurisprudence, theology, moral and natural philosophy* (Aldershot 2008) 13–28.
Wilson, Christopher, *The Gothic cathedral. The architecture of the great church 1120–1530* (London 2000).
Wingens, Marc, *Over de grens De bedevaart van katholieke Nederlanders in de zeventiende en achttiende eeuw* (Nijmegen 1994).
Wiskerke, Cornelis, *De afschaffing der gilden in Nederland* (Amsterdam 1938).
Wolff, Wilhelm Daniel, *Die Säkularisierung und Verwendung der Stifts- und Klostergüter in Hessen-Kassel unter Philipp dem Grossmütigen und Wilhelm IV. Ein Beitrag zur deutschen Reformationsgeschichte* (Gotha 1913).
Wollasch, Joachim, 'Monasticism: The first wave of reform', in: Timothy Reuter (ed.), *The New Cambridge Medieval History*, vol. III (Cambridge 1999) 163–185.
Woodside, Alexander, 'Some Mid-Ching theorists of popular schools: Their innovations, inhibitions and attitudes toward the poor', *Modern China*, 9 (1983) 3–35.
——, 'The divorce between the political center and educational creativity in Late Imperial China', in: Benjamin A. Elman and Alexander Woodside (eds.), *Education and society in Late Imperial China, 1600–1900* (Berkeley 1994) 458–492.
Woodside, Alexander and Benjamin A. Elman, 'Introduction', in: Benjamin A. Elman and Alexander Woodside (eds.), *Education and society in Late Imperial China, 1600–1900* (Berkeley 1994) 1–15.
——, 'Afterword. The expansion of education in Ching China', in: Elman and Woodside (eds.), Benjamin A. Elman and Alexander Woodside (eds.), *Education and society in Late Imperial China, 1600–1900* (Berkeley 1994) 525–560.
Woude, Ad van der (ed.), 'De Goldberg-enquête in het Departement van Texel, 1801', *AAG Bijdragen* 18 (1973) 95–250.
Wriedt, Markus, 'Continuity and competition: Luther's call for educational reform in the light of medieval precedents', in: Beat A. Kümin (ed.), *Reformations old and new. Essays on the socio-economic impact of religious change c.1470–1630* (Aldershot 1996).

Wrigley, Edward Anthony, *Continuity, chance and change. The character of the industrial revolution in England* (Cambridge 1988).

Wu, Pei-yi, 'Education of children in the Sung', in: De Bary, Wm.Theodore and John W. Chaffee (eds.), *Neo-Confucian education: The formative stage* (Berkeley 1989) 307–324.

Xiao-Wu, Guan, 'An interpretation of a *Shui Lun Fu* (Rhapsody on waterwheel)', in: Hong-Sen Yan and Marco Ceccarelli (eds.), *International symposium on History of machines and mechanisms Proceedings of HMM 2008* (s.l 2009) 215–224.

Yang, Ching Kun, *Religion in Chinese society* (Berkeley/Los Angeles 1961).

Yoshida, Tora, *Salt production techniques in Ancient China: the Aobo Tu* (tr. and rev. by Hans Ulrich Vogel, Leiden 1993).

Yü, Chün-fang, *The renewal of Buddhism in China: Chu-hung and the Late Ming synthesis* (New York 1981).

——, 'Ch'an education in the Sung: Ideals and procedures', in: Wm.Theodore De Bary and John W. Chaffee (eds.), *Neo-Confucian education: The formative stage* (Berkeley 1989) 57–104.

Yuan, Tsing, 'The porcelain industry at Ching-te-chen', *Ming Studies* 6 (1978) 45–53.

Zanden, Jan Luiten van, *The long road to the Industrial Revolution. The European economy in a global perspective, 1000–1800* (Leiden 2009).

Zanden, Jan Luiten van and Tine de Moor, 'Van fouten kan je leren. Een kritische benadering van de mogelijkheden van "leeftijdstapelen" voor sociaal-economisch onderzoek naar gecijferdheid in het pre-industrieel Vlaanderen en Nederland', *Tijdschrift voor Sociale en Economische Geschiedenis*, 5 (2008) 55–86.

Zanden, Jan Luiten van, 'Skill premium and the "Great Divergence"', *European Review of Economic History*, 13 (2009) 121–153.

Zeischka, Siger, *Minerva in de polder. Waterstaat en techniek in het hoogheemraadschap van Rijnland (1500–1856)* (Amsterdam 2007).

Zen-Sun, E-tu and Shiou-Chuan Su (tr.), *Sung Ying-Hsing: Chinese technology in the seventeenth century* (Dover 1966).

Zhang, Biachun, 'Ancient Chinese windmills', in: Hong-Sen Yan and Marco Ceccarelli (eds.), *International symposium on History of machines and mechanisms Proceedings of HMM 2008* (s.l. 2009) 203–214.

Zhang, Ling, 'Changing with the Yellow River: An environmental history of Hebei, 1048–1128', *Harvard Journal of Asiatic Studies*, 69 (2009) nr.1, 1–36.

Zhang, Qiong, 'About God, demons, and miracles: The Jesuit discourse on the supernatural in Late Ming China', *Early Science and Medicine*, 4 (1999) 1–36.

Zschunke, Peter, *Konfession und Alltag in Oppenheim. Beiträge zur Geschichte von Bevölkerung und Gesellschaft einer gemischtkonfessionellen Kleinstadt in der frühen Neuzeit* (Wiesbaden 1984).

Zuidervaart, Huib J., *Van 'konstgenoten' en hemelse fenomenen. Nederlandse sterrenkunde in de achttiende eeuw* (Rotterdam 1999).

Zürcher, Erik, 'Buddhism and education in T'ang times', in: Wm.Theodore De Bary and John W. Chaffee (eds.), *Neo-Confucian education: The formative stage* (Berkeley 1989) 19–56.

Zurndorfer, Harriet T., *Change and continuity in Chinese local history: The development of Hui-Chou prefecture, 800–1800* (Leiden 1989).

——, *China bibliography: A research guide to reference works about China past and present* (Leiden s.a.).

——, 'China and science on the eve of the "Great Divergence" 1600–1800: A review of recent revisionist scholarship in western languages', *History of Technology*, 29 (2009) 81–101.

——, 'The resistant fibre. Cotton textiles in Imperial China', in: Giorgio Riello and Prasannan Parthasarathi (eds.), *The spinning world. A global history of cotton textiles, 1200–1850* (Oxford 2009) 42–62.

INDEX

www.ingramcontent.com/pod-product-compliance
Lightning Source LLC
LaVergne TN
LVHW010657300926
849161LV00005B/1144
9789004233881